COMPLETE GUIDE TO READING SCHEMATIC DIAGRAMS

SECOND EDITION

COMPLETE GUIDE TO READING SCHEMATIC DIAGRAMS

SECOND EDITION

John Douglas-Young

PARKER PUBLISHING COMPANY, INC. West Nyack, N.Y.

© 1979, by

PARKER PUBLISHING COMPANY, INC.

West Nyack, N.Y.

All rights reserved. No part of this book may be reproduced in any form or by any means, without permission in writing from the publisher.

Reward Edition . . . . . . November 1980

Library of Congress Cataloging in Publication Data

Douglas-Young, John
   Complete guide to reading schematic diagrams.

   Includes index.
   1. Electronics--Diagrams. I. Title.
TK7866.D68   1979   621.381'022'3   79-15502

Printed in the United States of America

PREFACE

Since publication of the First Edition of *Complete Guide to Reading Schematic Diagrams* there have been many far-reaching developments in electronics technology, especially in the use of solid-state and integrated circuits — now widespread and increasing at a rate undreamed of a few years ago. This brand-new edition therefore includes many important new circuits in addition to the standard ones so essential to technicians. Credit is also due to the many readers of the earlier book whose suggestions and contributions have been most helpful in updating and making a broad range of improvements in this edition.

The organization of subject matter that proved so successful has been further improved in this new edition. Each chapter, with the exception of the first, covers one of the main groups of circuits. First, you get the basic circuit, with the distinctive features emphasized to enable you to identify it quickly, so you'll know right away whether it's an amplifier, an oscillator, or something else. You'll then get a brief description of the circuit, which will help you decide whether it's the one you need to investigate further.

If it is, you'll be shown how to break this circuit down into its subcircuits, with an explanation of each component, what its function is, and how a defect in it might affect the operation of the whole.

To help you in your circuit analysis there is also a comprehensive Appendix, with abundant supplementary data. Here you'll find charts, conversion tables, equations, definitions, and, most valuable of all today, a complete listing of the new SI units of the metric system, which is making itself felt more in electronics than in any other field of engineering. There's everything you need to get the most out of each chapter, placed in an easy ready-reference section for convenience in looking things up and to keep the main text uncluttered and simple.

To help you become a more skilled troubleshooter, chapter 1 gives you the tried and trusted procedures used by professionals, and in addition the all-new Circuit Analyzer. This is an important analytical tool that steers you to the right part of the book where you can find specimen circuit diagrams and descriptions to compare with the circuit you're examining, to confirm your identification, to find out how it works, and to diagnose its troubles, if any.

You'll find the Circuit Analyzer invaluable while building up your proficiency in reading circuit diagrams. Eventually, by using this ingenious device, you'll become so at home with schematics you'll read them as easily as a road map. This feature will save you much time and trouble in the early stages, and it won't be long before you're enjoying the study of these electrical "pictures," every one of which is worth the proverbial thousand words. You'll take pride in understanding them (and explaining them to others), and you'll be able to use them with confidence to troubleshoot and make faster repairs.

Efficiency in troubleshooting is one of the most valuable accomplishments you can have. You are the most important person around when things go wrong. You are the backbone of the service shop, maintenance crew, production line or electronics laboratory. And the most important tool in your toolbox is the schematic diagram.

John Douglas-Young

CONTENTS

PREFACE TO THE SECOND EDITION 5

1. SCHEMATIC DIAGRAMS 17

   The Purpose of a Schematic Diagram 17
   Graphic Symbols 17
   Class Letters 17
   Block Diagram 34
   Schematic Analysis 34
   Active and Passive Elements 38
   External Power 38
   Circuit Identification 38
   Circuit Analyzer 39
   How to Use the Circuit Analyzer 39
   Professional Troubleshooting 42
   General Troubleshooting Technique 43
   Visual Inspection 45
   Confirm Complaint 45
   Signal Tracing 46
   Waveform Analysis 48
   Signal Injection 48
   Brute Force 49
   Part Substitution 49
   D.C. Voltage Analysis 50
   Resistance Checks 50

2. VOLTAGE AMPLIFIERS 52

RE-COUPLED AUDIO VOLTAGE AMPLIFIERS 53
Distinguishing Features . Uses . Detailed Analysis . Circuit Variations

DIRECT-COUPLED AUDIO VOLTAGE AMPLIFIERS 61
Direct-Coupled Vacuum-Tube Audio Voltage Amplifier 62
Distinguishing Features . Uses . Detailed Analysis .
Direct-Coupled Transistor Audio Voltage Amplifiers 63
Distinguishing Features . Uses . Detailed Analysis . Circuit Variations

DIFFERENTIAL AMPLIFIERS 68
Distinguishing Features . Uses . Detailed Analysis . Feedback . Circuit Variations

OPERATIONAL AMPLIFIERS 72
Distinguishing Features . Uses . Detailed Analysis . Circuit Variations

VIDEO AMPLIFIERS 78
Distinguishing Features . Uses . Detailed Analysis . Circuit Variations

IF AMPLIFIERS 86
Distinguishing Features . Uses . Detailed Analysis . Circuit Variations

RF AMPLIFIERS 91
Distinguishing Features . Uses . Detailed Analysis . Circuit Variations

3. POWER AMPLIFIERS 96

SINGLE-ENDED AUDIO POWER AMPLIFIERS 96
Distinguishing Features . Detailed Analysis

PUSH-PULL AUDIO POWER AMPLIFIERS 99
Distinguishing Features . Uses . Detailed Analysis

INVERSE FEEDBACK   103
Circuit Variations

SINGLE-ENDED RADIO-FREQUENCY POWER AMPLIFIERS   107
Distinguishing Features . Uses . Detailed Analysis .
Circuit Variations

PUSH-PULL RF POWER AMPLIFIERS   110
Distinguishing Features . Uses . Detailed Analysis .
Circuit Variations

LINEAR RF AMPLIFIERS   112
Distinguishing Features . Uses . Detailed Analysis .
Circuit Variations

MAGNETIC AMPLIFIERS   113
Distinguishing Features . Uses . Detailed Analysis

4. OSCILLATORS **117**

L-C OSCILLATORS   117

HARTLEY OSCILLATOR   117
Distinguishing Features . Uses . Detailed Analysis

COLPITTS OSCILLATOR   120
Distinguishing Features . Uses . Detailed Analysis

ELECTRON-COUPLED OSCILLATOR   122
Distinguishing Features . Uses . Detailed Analysis

TUNED-PLATE, TUNED-GRID OSCILLATOR   123
Distinguishing Features . Uses . Detailed Analysis

CRYSTAL OSCILLATORS   125
Distinguishing Features . Uses . Detailed Analysis .
Circuit Variations

ARMSTRONG, OR TUNED-GRID OSCILLATOR   127
Distinguishing Features . Uses . Detailed Analysis

R-C (RELAXATION) OSCILLATORS 128

BLOCKING OSCILLATOR 128
Distinguishing Features . Uses . Detailed Analysis . Circuit Variations

ASTABLE OR FREE-RUNNING MULTIVIBRATOR 132
Distinguishing Features . Uses . Detailed Analysis . Circuit Variations

SAWTOOTH GENERATOR 135
Sync Input . Circuit Variations

SIMPLE RELAXATION OSCILLATOR 137
Distinguishing Features . Uses . Detailed Analysis . Circuit Variations

5. MODULATION **140**

AMPLITUDE MODULATION 140

PLATE MODULATION 141
Distinguishing Features . Uses . Detailed Analysis

GRID MODULATION 142
Distinguishing Features . Uses . Detailed Analysis

FREQUENCY MODULATION 144

REACTANCE CIRCUIT 144
Distinguishing Features . Uses . Detailed Analysis . Circuit Variations

ARMSTRONG BALANCED MODULATOR 147
Distinguishing Features . Uses . Detailed Analysis

PRE-EMPHASIS CIRCUIT 150

PHASE MODULATION 151

VECTORS 151

COLOR MODULATION 151

Contents

DOUBLY BALANCED MODULATOR 153
Distinguishing Features . Uses . Detailed Analysis

PULSE MODULATION 156
Uses . Circuits

MIXER CIRCUIT 158
*Distinguishing Features . Uses . Detailed Analysis .
Circuit Variations*

CONVERTERS 160
*Distinguishing Features . Uses . Detailed Analysis .
Circuit Variations*

6. DEMODULATION 166

DIODE DEMODULATOR - CRYSTAL DETECTOR 166
Distinguishing Features . Uses . Detailed Analysis

Automatic Volume Control 168
Circuit Variations

Automatic Gain Control 169
Peak Detector 170

VACUUM-TUBE DIODE DETECTOR 170
Distinguishing Features . Uses . Detailed Analysis

KEYED AGC 172
*Distinguishing Features . Uses . Detailed Analysis .
Circuit Variations*

GRID-LEAK DETECTOR 174
*Distinguishing Features . Uses . Detailed Analysis .
Circuit Variations*

REGENERATIVE DETECTOR 176
*Distinguishing Features . Uses . Detailed Analysis .
Circuit Variations*

FM DISCRIMINATOR 178
*Distinguishing Features . Uses . Detailed Analysis .
Circuit Variations*

RATIO DETECTOR 180
Distinguishing Features . Uses . Detailed Analysis . Circuit Variations

GATED-BEAM FM DETECTOR 181
Distinguishing Features . Uses . Detailed Analysis . Circuit Variations

ENVELOPE DETECTOR 184
Distinguishing Features . Uses . Detailed Analysis . Circuit Variations

DE-EMPHASIS 187

SYNCHRONOUS DETECTOR 187
Distinguishing Features . Uses . Detailed Analysis . Circuit Variations

7. POWER SUPPLIES 191

BATTERY POWER SUPPLY 191

UNIVERSAL AC-DC POWER SUPPLY 192
Distinguishing Features . Detailed Analysis . Circuit Variations

FULL-WAVE RECTIFIER 195
Distinguishing Features . Uses . Detailed Analysis

SOLID-STATE FULL-WAVE POWER SUPPLY 198
Circuit Variations

TRANSFORMERLESS VOLTAGE-DOUBLER POWER SUPPLY 198
Distinguishing Features . Uses . Detailed Analysis . Circuit Variations

TRANSFORMERLESS VOLTAGE-TRIPLER POWER SUPPLY 200

BRIDGE RECTIFIER POWER SUPPLY 201
Distinguishing Features . Uses . Detailed Analysis

VOLTAGE DIVIDERS 202

VOLTAGE REGULATION 203
Circuit Variations

HIGH-VOLTAGE POWER SUPPLIES 207

HIGH-VOLTAGE POWER SUPPLY FOR TRANSMITTER 207

HIGH-VOLTAGE POWER SUPPLY FOR CATHODE-RAY TUBE 208
Distinguishing Features. Uses. Detailed Analysis. Circuit Variations

TV HIGH-VOLTAGE SUPPLY 211
Distinguishing Features. Uses. Detailed Analysis. Circuit Variations

DC INVERTERS 215
Distinguishing Features. Uses. Detailed Analysis. Circuit Variations

Thyratrons as Industrial Power Controls 218
Diacs and Triacs 219
Triac Power-Control Circuit 219

8. MODIFIERS 221

Attenuators and Pads 221
Filters 223
Integration and Differentiation 229

9. LOGIC CIRCUITS 234

Inverter 236
AND Gate 236
OR Gate 237
RTL NOR Gate 239
DTL NAND Gate 241

TTL NAND Gate 242
AND and OR Gates from NAND and NOR 243
Flip-Flop 244

10. BRIDGE CIRCUITS 248

Wheatstone Bridge 248
LC Bridge 249
Owen Bridge 251
Maxwell Bridge 251
Hay Bridge 251
Resonance Bridge 252
Schering Bridge 252
Wien Bridge 252

11. SYNCHRO SYSTEMS 253

Servomechanisms 253
Synchro Systems 253
Synchro Schematics 255
Resolver 256
Amplidyne 259

12. OTHER CIRCUITS 250

MULTIVIBRATORS 260

MONOSTABLE MULTIVIBRATORS 261
 Distinguishing Features . Uses . Detailed Analysis

SCHMITT TRIGGER 263

BISTABLE MULTIVIBRATOR, OR FLIP-FLOP 263
 Distinguishing Features . Uses . Detailed Analysis

SAWTOOTH GENERATOR 266
 Distinguishing Features . Uses . Detailed Analysis

FREQUENCY DIVIDER 267
 Distinguishing Features . Uses . Detailed Analysis .
 Circuit Variations

FREQUENCY MULTIPLICATION 269
Distinguishing Features . Uses . Detailed Circuit Analysis

FREQUENCY COMPARATOR 270
Distinguishing Features . Uses . Detailed Analysis

PHOTOCELLS 272

CATHODE-FOLLOWER CIRCUIT 273
Distinguishing Features . Uses . Detailed Analysis . Circuit Variations

INHIBITING CIRCUIT 275

APPENDIX 277

Amplifier Classification 277
Transistor Operation 279
Color Coding 283
Preferred Values 285
International System of Units (SI Units) 286
Conversion Factors (U.S. and Metric) 289
Ohm's Law Formulas 290
Common Electrical Formulas 291
Greek Alphabet 292
Radio Frequency Spectrum 293

INDEX **295**

1
SCHEMATIC DIAGRAMS

The Purpose of a Schematic Diagram

A schematic diagram shows the electrical connections of an electronic device, using symbols and straight lines to represent the parts and their connections.

Graphic Symbols

A graphic symbol represents the *function* of a part in a circuit, not its outward appearance. Table I gives a list of graphic symbols in general use, and the class letters of the principal types.

Class Letters

As shown in Table I, the principal types of symbols have class letters. In a schematic all the components with the same class letter are numbered in sequence from left to right, and downwards where two or more are equidistant from the left-hand margin, for ease of correlation with descriptions, instructions and parts lists. Symbols which represent different sections of a multiple component, such as a dual-triode vacuum tube, or a ganged control in which one knob moves two or more decks or sections, are given the same number, but

TABLE I

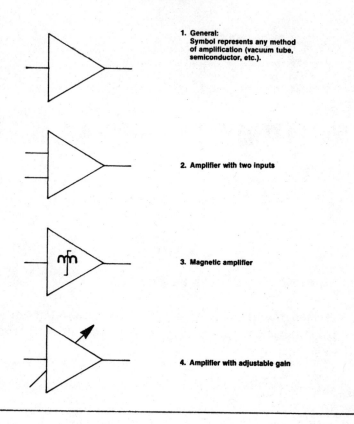

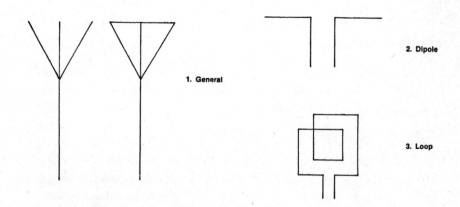

TABLE I (continued)

AUDIBLE SIGNALING DEVICE (LS)

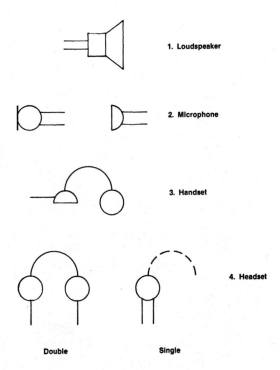

1. Loudspeaker

2. Microphone

3. Handset

4. Headset

Double   Single

BATTERY (BT)

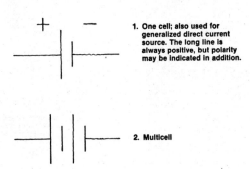

1. One cell; also used for generalized direct current source. The long line is always positive, but polarity may be indicated in addition.

2. Multicell

TABLE I (continued)

CABLE, CONDUCTOR, WIRING (W)

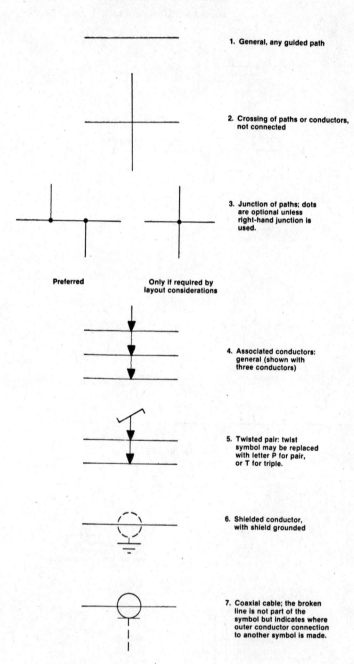

TABLE I (continued)

 8. Circular waveguide

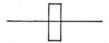 9. Rectangular waveguide

CAPACITOR (C)

Style 1   Style 2

1. Capacitor; the curved or modified electrode indicates the outside, low potential or movable element.

Style 1 modified to identify electrode

   2. Polarized capacitor

Style 1   Style 2

3. Adjustable or variable capacitor. If mechanical linkage of more than one unit is to be shown, the tails of the arrows are joined by a dashed line.

Style 1   Style 2

TABLE I (continued)

CIRCUIT BREAKER (CB)

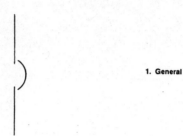

1. General

CONNECTOR: FEMALE (J), MALE (P)

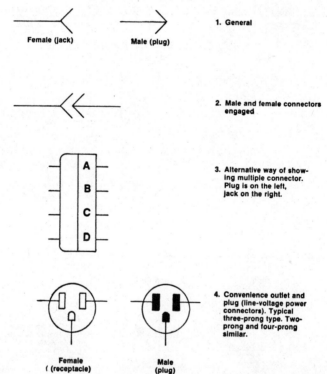

TABLE I (continued)

CRYSTAL UNIT, PIEZOELECTRIC (Y)

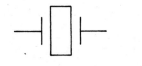

1. Piezoelectric crystal unit, including quartz crystal.

FUSE (F)

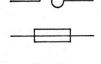

1. General: all three symbols are used.

GROUND, CIRCUIT RETURN (no class letter)

1. General: either earth, body of water, or chassis at zero potential

2. Chassis ground: may be at substantial potential with respect to earth ground.

INDUCTOR (L)

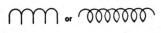

1. General: right-hand symbol is deprecated and should not be used on new schematics.

2. Magnetic core inductor

TABLE I (continued)

3. Tapped inductor

4. Adjustable inductor

INTEGRATED CIRCUIT (U)

1. General: unused pin connections need not be shown.

   The asterisk is not part of the symbol. It indicates where the type number is placed.

LAMP (DS)

1. General; light source, general

2. Glow lamp, neon lamp (a.c. type)

3. Incandescent lamp

4. Indicating, pilot, signaling or switchboard light

Schematic Diagrams

TABLE I (continued)

METER (M)

1. The asterisk is not part of the symbol. It indicates where to place a letter or letters indicating the type of meter:

   A = ammeter   W = wattmeter
   DB, VU = audio level meter
   F = frequency meter
   MA = milliammeter
   OHM = ohmmeter   V = voltmeter

RELAY (K)

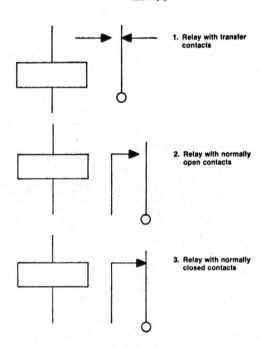

1. Relay with transfer contacts

2. Relay with normally open contacts

3. Relay with normally closed contacts

RESISTOR (R)

1. General

2. Tapped resistor

TABLE I (continued)

 3. Resistor with adjustable contact (potentiometer)

 4. Continuously adjustable resistor (rheostat)

 5. Thermistor

 6. Photoconductive transducer (e.g., cadmium-sulfide photocell)

SEMICONDUCTOR DEVICE
DIODE (D OR CR)

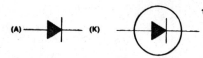 1. Semiconductor diode; enclosure symbol may be omitted where confusion would not be caused.
A = anode, K = cathode. The letters are not part of the symbol.

 2. Breakdown diode (zener diode)

 3. Tunnel diode

 4. Photosensitive diode

 5. Photoemissive diode, light-emitting diode (LED)

Schematic Diagrams 27

TABLE I (continued)

SEMICONDUCTOR DEVICES
(continued)

6. Thyristor or silicon controlled rectifier (SCR)

7. Triac
8. Diac (same as triac, but no gate lead)

TRANSISTOR (Q)

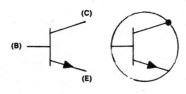
(B)
(C)
(E)

1. NPN transistor; enclosure symbol may be omitted where confusion would not be caused (unless an electrode is connected to it, as shown here).

 C = collector, E = emitter, B = base; these letters are not part of the symbol.

2. PNP transistor

3. NPN transistor with multiple emitters (four shown in this example)

(E)
(B2)
(B1)

4. Unijunction transistor with N-type base. If arrow on emitter points in opposite direction base is P type.

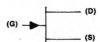
(G)
(D)
(S)

5. Junction field-effect transistor (JFET) with N-channel junction gate.

 G = gate, D = drain, S = source; these letters are not part of the symbol.

(G)
(D)
(S)

6. Insulated-gate field-effect transistor (IGFET) with N-channel (depletion type), single gate, positive substrate.

TABLE I (continued)

SEMICONDUCTOR DEVICES
(continued)

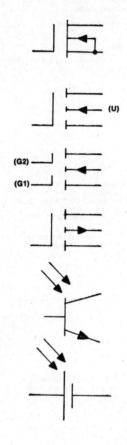

7. Insulated-gate field-effect transistor (IGFET), with N-channel (depletion type), single gate, active substrate internally terminated to source.

8. Insulated-gate field-effect transistor (IGFET) with N-channel (enhancement type), single gate, active substrate externally terminated.

    U = substrate; this letter is not part of symbol.

9. Same as previous example, but with two gates

10. Insulated-gate field-effect transistor (IGFET), with P-channel (enhancement type), single gate, active substrate externally terminated

11. Phototransistor (NPN type)

12. Photovoltaic transducer; barrier photocell; solar cell
    (No class designation letter)

SWITCH (S)

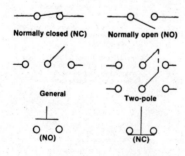

1. Single-throw switch; terminals are necessary for clarity in an NC switch, but may be omitted in an NO switch.

2. Double-throw switch

3. Push button

Schematic Diagrams

TABLE I (continued)

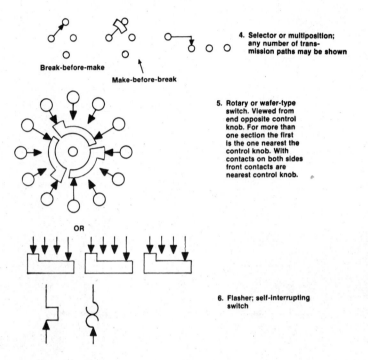

4. Selector or multiposition; any number of transmission paths may be shown

5. Rotary or wafer-type switch. Viewed from end opposite control knob. For more than one section the first is the one nearest the control knob. With contacts on both sides front contacts are nearest control knob.

6. Flasher; self-interrupting switch

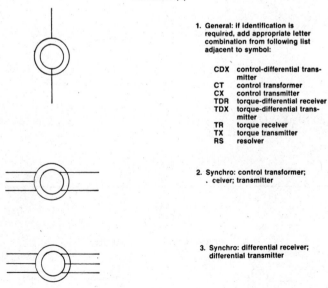

SYNCHRO (B)

1. General: if identification is required, add appropriate letter combination from following list adjacent to symbol:

   CDX   control-differential transmitter
   CT   control transformer
   CX   control transmitter
   TDR   torque-differential receiver
   TDX   torque-differential transmitter
   TR   torque receiver
   TX   torque transmitter
   RS   resolver

2. Synchro: control transformer; ceiver; transmitter

3. Synchro: differential receiver; differential transmitter

TABLE I (continued)

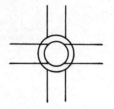

4. Synchro: resolver

PICKUP HEAD (PU)

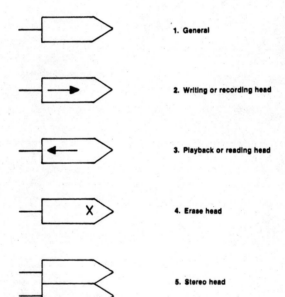

1. General

2. Writing or recording head

3. Playback or reading head

4. Erase head

5. Stereo head

PIEZOELECTRIC CRYSTAL UNIT (Y)

1. Also called quartz crystal unit

TABLE I (continued)

TRANSFORMER (T)

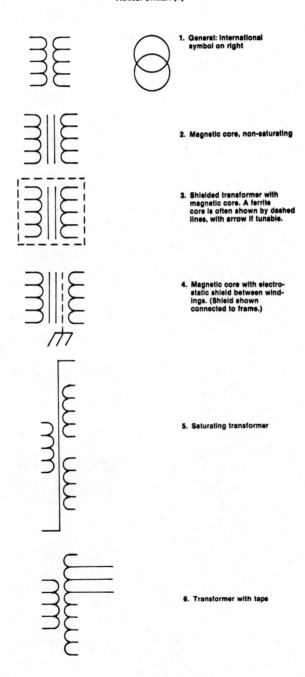

1. General: International symbol on right

2. Magnetic core, non-saturating

3. Shielded transformer with magnetic core. A ferrite core is often shown by dashed lines, with arrow if tunable.

4. Magnetic core with electrostatic shield between windings. (Shield shown connected to frame.)

5. Saturating transformer

6. Transformer with taps

TABLE I (continued)

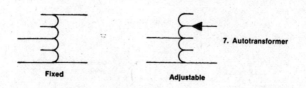

Fixed   Adjustable   7. Autotransformer

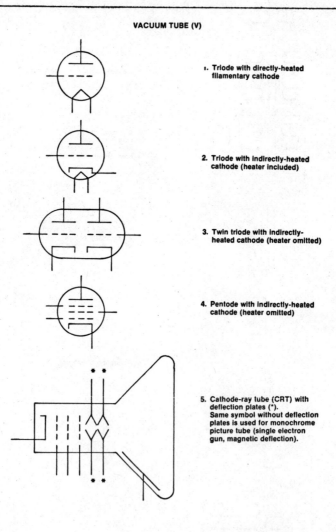

VACUUM TUBE (V)

1. Triode with directly-heated filamentary cathode

2. Triode with indirectly-heated cathode (heater included)

3. Twin triode with indirectly-heated cathode (heater omitted)

4. Pentode with indirectly-heated cathode (heater omitted)

5. Cathode-ray tube (CRT) with deflection plates (*). Same symbol without deflection plates is used for monochrome picture tube (single electron gun, magnetic deflection).

TABLE I (continued)

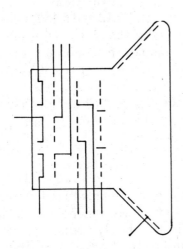

6. Color picture tube with three electron guns and electromagnetic deflection

7. Cold-cathode gas-filled tube

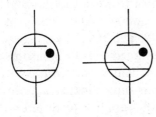

8. Mercury-pool vapor-rectifier tubes; right-hand tube has ignitor.

9. Thyratron with indirectly-heated cathode (heater shown)

the individual sections are identified by letter suffixes. For example, C2A, C2B and C2C would designate three sections of a ganged variable capacitor (see Figure 2.36, 5.11, 5.12 and 5.15 for examples).

Where a schematic diagram is broken down into subordinate sections because it is too large or complex to be shown on one sheet of paper, it is customary to use three-figure number groups starting with 1, 2 and so on, for each section (like the numbering of rooms by floors in a hotel). For example, C102, R115 and Q167 will all be in the first section, and R455, C407 and Q410 will all be in the fourth.

Block Diagram

Just as a book starts with a table of contents to tell you where to locate the different chapters, and what they are about, so a large and complex schematic diagram is frequently "summarized" in an accompanying *block diagram*. This is a diagram in which the essential units of the overall system are drawn in the form of blocks, and their relationship to each other is indicated by connecting lines. These lines usually represent the signal paths, and often have arrows to show direction. A typical block diagram is reproduced in Figure 1.1.

A block diagram is useful because it gives you a quick general view of the overall system. Where one is provided, you can use it to identify the circuit because the block names it. Then you can look for it in the index of this book to find what page to turn to for a full description.

When such a block diagram is not available, you can use the universal block diagram in Figure 1.2. This take advantage of the fact that most electronic equipment is designed to act upon some input so as to obtain a desired output. Consequently, the sequence of circuits will be much the same in all of them, as shown in the diagram. Table II gives you some examples of how this applies to different types of electronic equipment. You can construct your own block diagram by copying Figure 1.2, substituting specific designations for the general ones in the figure.

Schematic Analysis

Schematic analysis consists in separating the overall electrical picture into its parts, so as to determine their nature, function and

Schematic Diagrams

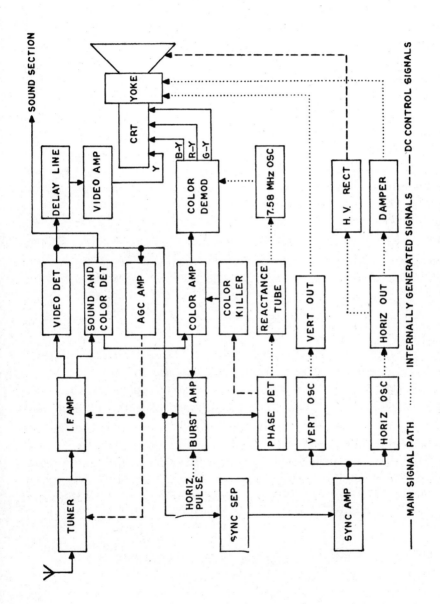

Figure 1.1 TV Color Set Block Diagram

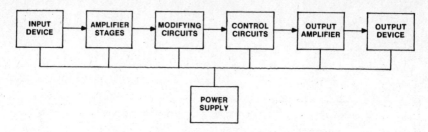

Figure 1.2   Universal Block Diagram

TABLE II — APPLICATION OF UNIVERSAL BLOCK DIAGRAM

TYPE OF EQUIPMENT	INPUT DEVICE	AMPLIFIER STAGES	MODIFYING CIRCUITS	CONTROL CIRCUITS	OUTPUT AMPLIFIER	OUTPUT DEVICE
Public Address	Microphone	Preamplifier	Tone Controls	Gain Control	Power Amplifier	Speaker
Radio Receiver	Antenna, R.F. Front End	I.F. Amplifier	Second Detector	Volume Control	Audio Section	Speaker
Radio Transmitter	Crystal Oscillator	Buffer Amplifier	Modulator	Impedance Matching Network	Final Amplifier	Antenna
Video channel of TV Receiver	Antenna, Tuner	Video I.F. Amplifier	Video Detector	Brightness, Contrast Controls	Video Amplifier	Picture Tube

relationship. It is a process like anatomy. The following chapters will tell you about the "organs" found in schematic "bodies."

These "organs" are individual circuits. Taken separately, you'll find each is easy to understand and recognize, because all circuits are laid out in a conventional way. The signal input is almost invariably from the left, the output to the right. This horizontal flow takes place along two parallel paths (see Figure 1.3). The upper path is the "high" side of the circuit, the lower is the "low." The low side is the return path, and is usually considered to be at zero or ground potential, even though it may not be connected to a real earth ground at all.

In some diagrams the return path or low side of the circuit is emphasized with a heavier line (as in Figure 1.4). But in many others it is left to the imagination, and connections to it are indicated by use of the circuit return, or ground symbol, shown in Table I and Figure 1.5. This is understandably necessary in complicated schematics, especially in small-scale reproductions. In this book, however, we shall generally

Schematic Diagrams

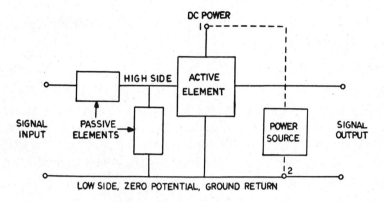

Figure 1.3   Circuit Conventions

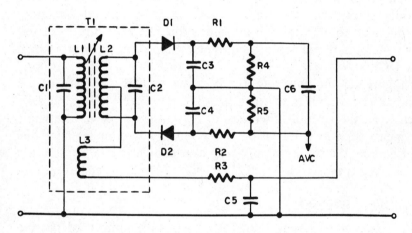

Figure 1.4   Typical Circuit Schematic
(Parts Values Omitted)

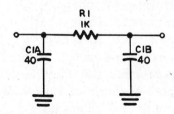

Figure 1.5   Pi Filter
(Parts Values Included)

stick to showing the complete return path, as we are dealing mostly with individual circuits.

Active and Passive Elements

Circuit elements are either active or passive. Passive elements are those which do not require an external source of power in order to function. They include resistors, capacitors, inductors and the like. Active elements are vacuum tubes, transistors and other devices that use external power, which they introduce into the circuit to modify an existing signal or to generate a new one.

External Power

When an active element is included in a circuit the external power is generally shown as entering and leaving in a vertical direction, perpendicular to the signal flow. Since this power is direct current (DC), its path is also characterized by the absence of capacitors (which block DC).

In Figure 1.3 the power source is drawn as if it were part of the individual circuit, but the dashed lines indicate that this is not usually the case. In the overall schematic of a complex piece of electronic equipment, such as a television set, where many circuits share one power source, the power supply would be shown separately, and the external connections to it (1 and 2 in Figure 1.3) might even be omitted.

Circuit Identification

The majority of circuits consist of a small number of passive elements built around an active element. The type of active element gives you your first hint as to the nature of the circuit. A tube or transistor has a type number, apart from the reference number, and if you look it up in a tube or transistor manual you'll often see what type of circuit it is likely to be used in. Some tubes, for instance, are only used in certain types of circuit. A 12BE6, which is a pentagrid converter, is only used in a converter circuit, so you know what type of circuit it is without looking further. But this is like looking up words in the dictionary. The purpose of this book is to show you how to read schematics without having to look things up.

Circuit Analyzer

Until you have had some practice, however, you will find it helpful to use the Circuit Analyzer as a guide in identifying circuits. This is a

Schematic Diagrams

means to steer you to the right part of the book, where you'll find specimen circuit diagrams and descriptions which you can compare with the circuit you're interested in, to confirm your identification, find out how it works, and diagnose troubles in it if necessary.

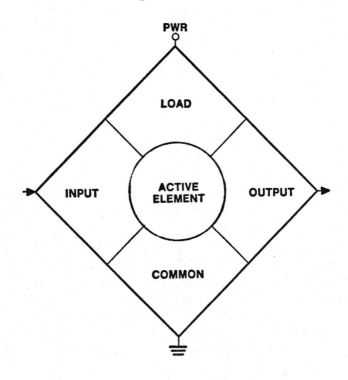

ACTIVE ELEMENT	PASSIVE ELEMENTS			
	INPUT	OUTPUT	COMMON	LOAD

Figure 1.6   Circuit Analyzer

How to Use the Circuit Analyzer

The Circuit Analyzer is shown in Figure 1.6. To use it you perform the following steps:

(1) Make a table similar to that beneath the diamond-shaped outline in Figure 1.6.
(2) Mentally superimpose the diamond-shaped outline of Figure

1.6 on the circuit you are examining, so that the active element is within the circle.
(3) Look in Table I for the symbol of the active element.
(4) Write down its class letter, and the number preceding the corresponding paragraph in the right-hand column of Table I, in the first box from the left of the table you made in step (1), under ACTIVE ELEMENT. If there is only one active element, write 00 in the second box.
(5) Now look at the part of the circuit that lies in the INPUT area of the Circuit Analyzer outline that you're still mentally superimposing over the circuit, and select the two principal passive elements in it.
(6) Look in Table I for their symbols.
(7) Write down their class letters and numbers (as you did those of the active element) in the third and fourth boxes, under INPUT. If there is only one passive element, enter 00 in the fourth box. If there are no passive elements in the input, write 00 in both boxes.
(8) Repeat steps (4), (5) and (6) for the OUTPUT, COMMON and LOAD areas.
(9) Now turn to Table III, and look for a combination of letters and numbers the same as, or as close as possible to, the combination you have just written down.
(10) Turn to the corresponding figure given in the table, and compare it with your circuit. Read the accompanying text to complete your analysis.

For example, suppose you're looking at the circuit shown in Figure 2.1 at the beginning of the following chapter. Figure 1.7 gives it with the outline of Figure 1.6 superimposed. The active element V1 is in the circle and the various passive elements are located in the appropriate areas.

The active element is a single vacuum-tube triode. Its symbol in Table I has the class letter V and the number 2, so we must enter V2 00 under ACTIVE ELEMENT.

In the INPUT area we have a fixed capacitor C1 and a fixed resistor R1. From Table I we determine that C1 R1 is the correct entry to go under INPUT. In the OUTPUT area there is one capacitor C3, so the entry under OUTPUT is C1 00. Similarly, the COMMON area has a fixed resistor R3 and capacitor C2, while the LOAD area has one resistor R2. The entries under COMMON and LOAD are therefore R1 C1 and R1 00 respectively.

Schematic Diagrams 41

TABLE III — CIRCUIT LOCATOR

ACTIVE ELEMENT		PASSIVE ELEMENTS									
		INPUT		OUTPUT		COMMON		LOAD		FIGURE	TYPE OF CIRCUIT
00	00	C1	$D1^2$	$C1^2$	L2	00	00	00	00	7.5	Power supply
00	00	C1	$D1^3$	$C1^2$	L2	00	00	00	00	7.6	Power supply
00	00	T3	D1	L1	C1	00	00	00	00	6.2	Video detector
00	00	T3	D1	R3	C1	00	00	00	00	6.1	Demodulator
00	00	T3	$D1^2$	R1	C1	00	00	00	00	6.7,6.8	Demodulator (FM)
00	00	T3	$D1^4$	$C1^2$	R1	00	00	00	00	6.10	Demodulator (Stereo)
00	00	T4	$D1^4$	$C1^2$	L2	00	00	00	00	7.7	Power supply
00	00	T4	$D1^4$	$C1^2$	R1	00	00	00	00	7.4	Power supply
A2	00	R1	00	00	00	R1	00	R1	00	2.19 (b)	Operational amplifier
A2	00	T3	C1	T3	C1	00	00	00	00	2.34	TV-PIX-IF system
Q1	00	00	00	00	00	00	00	R1	00	2.13,2.14	Audio voltage amplifier
Q1	00	00	00	00	00	R1	00	R1	00	2.13,2.17,2.19	Audio voltage amplifier
Q1	00	C1	R1	C1	00	R1	C1	R1	00	2.2	Audio voltage amplifier
Q1	00	C1	R1	L1	L1	R1	C1	R1	00	2.22	Video amplifier
Q1	00	L1	C3	L1	00	00	00	L1	00	4.2	Oscillator
Q1	00	L1	C3	L1	00	00	00	L1	R1	4.4	Oscillator
Q1	00	L2	C3	L2	C3	R1	C1	R1	00	2.36	RF amplifier
Q1	00	T3	C1	T3	C1	R1	C1	R1	00	2.32	IF amplifier
Q1	00	Y1	C1	L1	00	00	00	R1	00	4.8	Oscillator
Q1	Q1	00	00	00	00	00	00	R1	00	2.15	Audio voltage amplifier
Q1	Q1	(X-connected)		00	00	R1	R1	9.13			Multivibrator
Q2	00	C1	R1	T2	LS1	R1	C1	00	00	3.2	Audio power output amplifier
Q2	00	T2	C1	T2	00	00	00	R1	00	4.11	Oscillator
Q2	Q2	00	00	C1	LS1	R1	R1	00	00	3.4	Audio power output amplifier
Q2	Q2	(X-connected)		R1	C1	R1	R1	4.15			Multivibrator
Q2	Q2	T2	00	T2	LS1	R1	R1	00	00	3.9	Audio power output amplifier
Q4	00	00	00	C1	00	R1	00	R1	C1	4.17	Oscillator
Q10	Q10	00	00	00	00	00	00	00	00	A.4	Audio voltage amplifier
V1	00	T1	C3	T1	C3	C1	T2	00	00	3.10	RF power amplifier
V1	V1	T1	C3	T1	C3	C1	T2	00	00	3.11	RF power amplifier
V2	00	C1	R1	C1	00	R1	C1	R1	00	2.1	Audio voltage amplifier
V2	00	L1	C3	L1	00	00	00	L1	00	4.1,4.3,4.9	Oscillator
V2	00	L1	C3	T1	C3	00	00	L1	00	4.6	Oscillator
V2	00	R1	00	00	00	R1	00	R1	00	2.18	Audio voltage amplifier
V2	00	T2	C1	T2	00	00	00	R1	00	4.10	Oscillator
V2	00	Y1	R1	T1	C3	00	00	L1	00	4.7	Oscillator
V2	V2	C1	R1	C1	C1	R1	C1	R1	R1	3.8	Phase splitter
V2	V2	L3	L4	L3	L4	C1	00	R1	00	2.35	RF cascode voltage amplifier
V2	V2	(X-connected)		00	00	R1	R1	4.12			Multivibrator
V2	V2	(X-connected)		R1	00	R1	R1	4.13			Multivibrator
V4	00	C1	R1	L1	L1	R1	R3	R1	00	2.21	Video amplifier
V4	00	C1	R1	T2	LS1	R1	00	00	00	3.1	Audio power output amplifier
V4	00	L1	C3	L1	00	00	00	00	00	4.5	Oscillator
V4	00	T1	C3	T1	C3	00	00	00	00	3.12	RF power amplifier
V4	00	T3	00	T3	00	R1	00	R1	00	2.33	Video IF amplifier
V4	00	T3	C1	L1	C1	R1	C1	R1	00	6.9	Demodulator (FM)
V4	00	T3	C1	T3	C1	R1	C1	00	00	2.31	IF amplifier
V4	V4	C1	R1	T2	LS1	R1	00	00	00	3.3	Audio power output amplifier

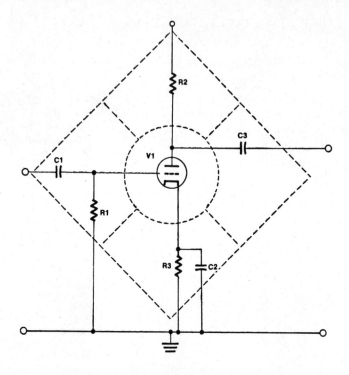

Figure 1.7 Using the Circuit Analyzer

When we look up this combination in Table III we are referred to Figure 2.1 for the circuit of an audio voltage amplifier, which is, of course, just what we would have expected.

Sometimes there are two active devices that operate as a pair. You enter them both, and this usually makes the circuit easier to identify.

Transistors may be n-p-n or p-n-p in circuits that are identical except for the power supply polarities. It is, therefore, perfectly correct to read Q2 for Q1, and vice versa, in most cases.

Among passive elements, always enter the most distinctive-looking ones, as they are the key ones in identifying a circuit.

Professional Troubleshooting

We are going to conclude this chapter by showing you how to troubleshoot any type of electronic equipment the way professionals do it. The experienced technician starts by getting the service manual, which contains the schematic diagram and much other valuable information. He lays it on the bench beside the equipment before he does anything else, because the information it contains, even on how to remove the cover, may save him time, and spare the equipment from the risk of unnecessary damage caused by "hit or miss" operations, especially if the equipment is new to him.

General Troubleshooting Technique

Many technicians hesitate to tackle repair jobs on electronic equipment with which they are not familiar, feeling that the task may prove to be beyond their skills and knowledge. The experienced "old timer," on the other hand, may take on such jobs without hesitation, for he realizes that *all* electronic equipment of a given general type works on the same principles, can suffer the same defects, and is amenable to the same troubleshooting techniques. In fact, if we consider complaints in broad rather than limited terms, specific defects may cause the same complaints in all types of equipment. The physical manifestations of the "complaint" may vary considerably from one type of equipment to another, however.

For instance, an amplifier can cause distortion due to a "leaky" transistor. This broad complaint (distortion) shows up in different ways according to what the amplifier is being used for:

(a) In the video section of a TV set, the distortion may show up as a *smeared picture*.
(b) In an audio amplifier or radio receiver, it may show up as *garbled sound*.
(c) In a radio transmitter, it may show up as *excessive harmonic radiation*.
(d) In a test instrument, it may show up as *inaccurate readings*.
(e) In an item of industrial control equipment, it may show up as *erratic operation*.

But in every case the defect is the same, the basic circuit is the same, the broad complaint is the same, and the defect can be found using the same general troubleshooting techniques (signal tracing, for example).

Thus it is possible to provide the chart given in Table IV, which applies to all types of electronic equipment. It is used to select a suitable diagnostic test technique. To use this chart, first determine the *general*

TABLE IV — UNIVERSAL TROUBLESHOOTING CHART

DIAGNOSTIC TECHNIQUE \ COMPLAINT	VISUAL INSPECTION	CHECK POWER SUPPLY	VOLTAGE CHECKS	SIGNAL TRACING	SIGNAL INJECTION	BRUTE FORCE	PARTS SUBSTITUTION	COMPONENT TESTS	ALIGNMENT	REMARKS
DEAD	X	X	X	X	X			X		
WEAK	X	X	X	X	X				X	Align only if other tests indicate
INTERMITTENT	X	X				X	X	X		
DISTORTION	X	X	X	X				X		"Weak" and "distorted" may be a common complaint
OSCILLATION	X	X		X			X		X	Similar tests for complaints of "hum" and "noise." Align only if other tests indicate

complaint. To do this you'll have to interpret the specific complaint in terms of its broad technical nature, as was done in the examples given above. For instance, if the equipment *doesn't work at all* it is DEAD, whether the item is a receiver, audio amplifier, industrial control, transmitter or test instrument. If the equipment *works, but lacks its usual response*, it is WEAK. In a radio transmitter this may show up as lowered power output, in a receiver as lack of sensitivity, in control equipment as sluggish operation. If the equipment *works "now and then"* or if some other complaint "comes and goes," it is considered INTERMITTENT. If the equipment *works, but not quite properly*, the complaint is DISTORTION, as in the examples given above. If the equipment is *unstable*, the trouble is OSCILLATION. The test methods given for this complaint also apply where circuit operation is upset by an undesired signal such as *hum, noise* or *interference*.

In most cases any of several test techniques may be used, depending on the test instruments available and the type of equipment serviced. Two general checks should be made regardless of complaint: *visual inspection* and *power supply verification* (incorrect supply voltages can cause a variety of problems).

Schematic Diagrams

Where alignment is indicated by the table, this applies only to equipment with fixed tuned circuits.

Visual Inspection

Table V outlines in block diagram form basic service procedure for all equipment. The first step, of course, is a quick visual inspection of the equipment, watching for obvious physical or electrical damage. This, in itself, may permit a quick isolation of the defect. At this time, check with the "customer" to see if the equipment has been subjected to unusual operating conditions, such as high temperature, humidity and so on.

TABLE V — BASIC STEPS IN SERVICING ALL TYPES OF EQUIPMENT

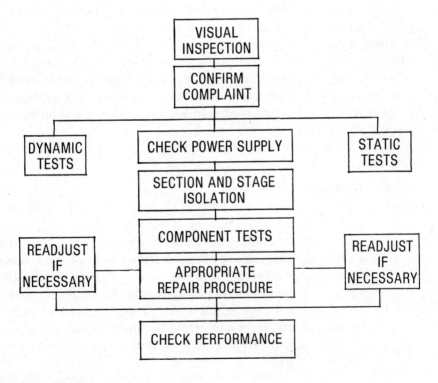

Confirm Complaint

Always *confirm the complaint*. This means, try out the equipment to make sure the "complaint" is as described by the "customer." Too often you may be told "it doesn't work" when what is meant is that it doesn't work as well as it should. A technician would use this expression only to describe a dead set.

After the initial visual inspection and check of power supply (which includes batteries, of course), you'll select the appropriate static or dynamic tests to isolate the trouble to the particular stage section.

Dynamic tests are tests made under actual operating conditions. These tests may be used to isolate all types of troubles, including not only serious component defects causing a major change in equipment operation, but also minor defects which affect circuit efficiency and performance, but permit "nearly normal" operation. Dynamic tests are also valuable for tracking down intermittent defects.

Signal Tracing

Signal tracing is the most powerful technique, involving tracing a signal stage by stage as it passes through the equipment. The instrument used to follow the signal may be an a.c. voltmeter, a signal tracer or an oscilloscope. Generally, a signal tracer is used to follow signals through receivers and audio amplifiers, while an oscilloscope is superior for most other items. An oscilloscope can also be used to check receivers if it is provided with a demodulator probe or has the necessary bandwidth.

To use the signal tracing technique, as illustrated in Figure 1.8, proceed as follows:

(a) Turn the equipment on.
(b) Apply an adequate test signal to the equipment's input. In a receiver this signal may be obtained from an r.f. signal generator, or by tuning in a local station. In an audio amplifier a suitable signal may be obtained from a record player, tuner or audio generator. In other cases, depending on the equipment being tested, the signal may be obtained from a pulse generator, square-wave generator, sawtooth oscillator or other device. If the equipment generates its own signal (for instance, a radio transmitter), a separate signal source may not be needed.
(c) Make sure that the connection of the signal source does not disturb the circuit. If necessary, insert a d.c. blocking capacitor, matching pad, or other device as necessary, in series with the input.

Schematic Diagrams 47

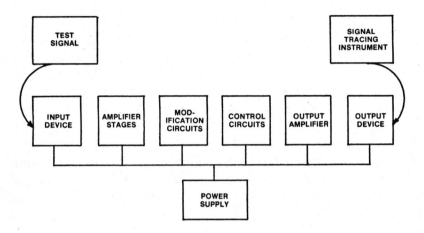

Figure 1.8 Signal Tracing Technique (see text). The test signal is applied where shown and the signal tracing instrument is connected to the input and output of each stage in turn. (This block diagram can be applied to almost any type of electronic equipment, as explained previously in Table II.)

(d) Make sure the test signal does not overload the equipment. Always use the minimum signal needed for usable output.
(e) Use a signal tracing instrument (signal tracer, oscilloscope, etc.) to check the relative amplitude and quality of the signal at the input and output of each stage, as shown in Figure 1.8. Depending on the instrument used, the signal may be: (1) heard in a loudspeaker; (2) observed as the closure of a "tuning eye"; (3) indicated by the deflection of a meter pointer; or (4) seen as a waveform on a CRT.
(f) The test signal should be modified by each stage. In the case of an amplifier, for instance, the signal should be increased in amplitude. In a clipper stage a portion of the signal should be removed or clipped off.
(g) If the signal is changed in an unexpected fashion (the amplitude drops instead of increasing, for instance), trouble is indicated and the defective stage has been isolated.
(h) Where necessary, change the type of pick-up (probe) as the signal is followed through the equipment. For example, an r.f. detector probe is used for checking the r.f. and i.f. stages of a receiver, a direct probe for the audio section.

Waveform Analysis

Waveform analysis is used to analyze changes in equipment performance that are minor rather than major or catastrophic—a deterioration in frequency response, for instance. Here, the oscilloscope is the instrument to use. The test signal may be a sine wave, square wave, pulse, AM or FM carrier, or other signal, depending on the nature of the test and the equipment being tested. To use this technique, apply a signal of accurately known characteristics and, using a 'scope, observe the signal waveform and the input and output of each stage. Changes in the signal waveform tell you a lot about stage operation. A "rounding" of a square wave, for example, indicates a falling off of an amplifier's high frequency response.

Signal Injection

Signal injection is another dynamic test technique. It is complementary to signal tracing, and is almost as powerful for tracking down trouble. As the name indicates, this technique involves injecting a test signal into the equipment:

    (a) Connect an indicator to the equipment's output stage. This may be an a.c. voltmeter, oscilloscope, or even the equipment's own output device (loudspeaker, TV screen, etc.).
    (b) Turn the equipment on.
    (c) Apply an appropriate test signal to the input of the *output* stage. The test signal may be obtained from an r.f. signal generator when checking the i.f. and r.f. stages of radio receivers, or from an audio generator when checking the audio stages of receivers, P.A. amplifiers, hi-fi and similar equipment; or from a pulse or square-wave generator when testing for frequency response and so on.
    (d) Make sure that the connection of the signal source does not change the normal operation of the stage by using a small coupling capacitor in series with the input lead if necessary.
    (e) Adjust the input signal level to the minimum signal needed for a normal output indication; later, readjust this signal level as necessary to prevent overload. Do this by adjusting the signal generator's attenuator control.

Schematic Diagrams

(f) Transfer the signal injection point *back*, stage by stage, from the equipment's output stage to its input point (A in Figure 1.8).
(g) Change to a different type of test signal where necessary: for example, an audio signal for the audio stages, a modulated r.f. signal of the proper frequency for the i.f. stages and converter stage, and a modulated r.f. signal for the r.f. and antenna stages.
(h) Note changes in equipment's output as shown by the output indicator. Normally, output level will go *up* as the signal injection point is transferred back, requiring readjustment of the signal generator's attenuator.
(i) Unexpected changes in the equipment's output signal indicate that the defective stage has been isolated. In the case of a "dead" stage, for example, the output signal will drop to zero when the injection point is shifted from the output to the input side of the stage.

Brute Force

Brute force testing means lightly tapping, wiggling, or otherwise manipulating the circuitry to make it exhibit an intermittent symptom. Electronic equipment that has been misbehaving in this manner usually is as good as gold as soon as it gets on the bench, and must be stimulated into bad performance. Some types of intermittent behavior are temperature related. In this case you may get it to act up by warming it in a box with a 40-watt light bulb as a heater, or spraying suspected parts with freon to cool them below their normal temperatures, depending upon what is required.

Part Substitution

Part substitution means, of course, replacing a suspected item with another known to be good, to see if that clears up the problem. Obviously, if it does, you have isolated the defect. This is a hangover from the old tube days, but still has a limited use in modern equipment, especially modular equipment, where a suspected module or circuit board can be easily removed, and a good one plugged in.

D.C. Voltage Analysis

D.C. voltage analysis is perhaps the oldest of all test techniques, and is valuable for isolating defects that cause a change in the equipment's operating voltages. The procedure is as follows:

(a) Turn the equipment on, and adjust any controls for normal operation.
(b) Using a d.c. voltmeter, check voltages between each active element pin and "ground," being careful to observe proper polarity.
(c) Compare the readings obtained with the service manual or schematic.
(d) Ignore *minor* discrepancies. *Major* differences indicate trouble.

In addition to d.c. voltage tests, *current measurements* are sometimes helpful for locating troubles in precision control devices and instruments. An overall current test may be made as a check on equipment efficiency. For this the meter is inserted in series with one of the power supply (or battery) leads.

Resistance Checks

Point-to-point *resistance checks* are useful for isolating defects which cause a change in d.c. resistance values. However, you need to remember that in equipment employing transistors misleading results can arise because of the direct resistive connection between a transistor's various electrodes. Since resistance tests apply only to the external circuit around the active element, use the following technique:

(a) Turn the equipment off.
(b) Discharge any large filter or bypass capacitors, using a short piece of wire or a clip lead.
(c) Remove active elements, if sockets are used; if soldered in place, remove them *only* from the stages to be checked.
(d) Using an ohmmeter, measure the d.c. resistance between each active element's connection points (each socket pin, for example, if sockets are used) and either circuit ground or common power connection.
(e) Compare the readings obtained with those given in the equipment's service manual or in the schematic diagram.

Schematic Diagrams

(f) Ignore *minor* differences between "actual" and "expected" values. *Major* differences indicate trouble.

If the foregoing procedures have not yet identified the circuit component causing the trouble you will have to test each one separately. But how much better it is to have to do this with only a few instead of checking every one of the hundreds that make up many pieces of electronic equipment!

The following chapters cover a very large number of individual circuits, explaining how they work as well as identifying them. Being able to read a schematic diagram is essential to understanding it, and knowing what the circuit is supposed to do when it is operating correctly is basic to the art of troubleshooting.

2
VOLTAGE AMPLIFIERS

A voltage amplifier is a means of using the current from a power supply or battery to produce a stronger replica of a weaker original signal. The weak signal is able to control the flow of current from the power source by means of a vacuum tube or transistor. The circuit, therefore, consists of three subcircuits: one for direct current from the power source, one for the weaker input signal, and one for the more powerful output signal.

In a voltage amplifier the emphasis is on the *voltage* of the output signal. Since current flows at all times this type of amplifier wastes power and its efficiency is low, but its output is free of distortion because the excursions of the input signal remain within the distortion-free limits of the tube or transistor.

Voltage amplifiers are used to amplify audio-frequency, intermediate-frequency, and radio-frequency signals. You can usually tell what frequency the circuit is designed to handle by the type of *coupling* that connects the input signal from the preceding circuit, and connects the output signal to the following circuit. The types of coupling most used for the three frequency ranges are as follows:

FREQUENCY RANGE OF AMPLIFIER	TYPE OF COUPLING
Audio frequency (AF)	Resistance-capacitance (RC) Direct
Intermediate frequency (IF)	IF transformer
Radio frequency (RF)	Impedance (inductance-capacitance)

Voltage Amplifiers 53

RC-COUPLED AUDIO VOLTAGE AMPLIFIERS

Figure 2.1 shows an audio voltage amplifier with a vacuum tube. Figure 2.2 shows the same circuit with a transistor. Ignore for the moment the extra resistor R1 in Figure 2.2. You'll see the reason for it later.

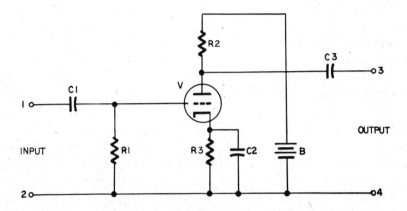

Figure 2.1 Audio Voltage Amplifier with Vacuum Tube

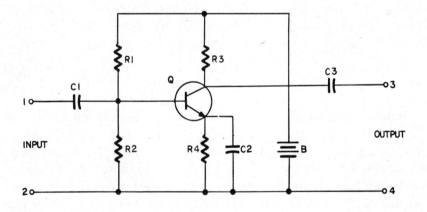

Figure 2.2 Audio Voltage Amplifier with Transistor

Distinguishing Features

These circuits clearly fit the description of an amplifier given above. In Figure 2.1 you see a circuit in which a weaker signal input is applied between the grid and cathode of a tube in order to control the flow of current from the battery so as to produce a stronger signal in the output circuit connected between the plate and cathode. Because the cathode is common to both input and output this is a common-cathode amplifier—more often called a *grounded-cathode* amplifier.

Similarly, in Figure 2.2 the emitter of the transistor is common to both input and output, so this is a *common-emitter* amplifier.

Tubes and transistors can be connected with either of their other two elements common (or grounded), but the grounded-cathode and common-emitter circuits are the most used because they give the highest gain.

Amplifiers must be *voltage* amplifiers or *current* amplifiers. The latter, which are more often called power amplifiers, must be able to pass a fairly heavy current. The presence of *load resistors* (R2 in Figure 2.1, R3 in Figure 2.2), which always have a resistance of several kilohms, will not allow a large current output, thus eliminating the possibility that these are power amplifiers. They are, therefore, *voltage amplifiers*.

The absence of any inductors in these circuits shows you that they are handling *audio* frequencies. AF circuits sometimes have iron-core transformers or chokes, but designers today generally avoid them wherever possible, to lower costs and to obtain a more even response to audio frequencies. Since iron-core inductors are the only type of inductor used at audio frequencies, the presence of any other type would indicate the circuit is used for higher frequencies.

Uses

Audio voltage amplifiers are used in radio and television, hi-fi, tape recorders, phonographs, public address systems and the like, to build up weak signals to the point where they can be used to control power amplifiers. For this reason you wouldn't expect to find an audio voltage amplifier in the final stage of a radio or stereo amplifier, but in the preceding stage or stages handling audio frequencies. Thus the position of the amplifier in the overall schematic is an additional way in which you can tell what type it is.

Detailed Analysis

DC Subcircuit: In Figure 2.3 electron flow is from B− through R3, V and R2, and back to B+. Although the power source is shown as a

Voltage Amplifiers

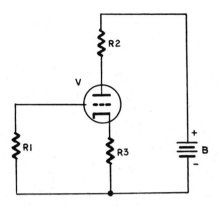

Figure 2.3 Vacuum-Tube DC Circuit

battery it is in practice more likely to be one of the electronic power supplies discussed in chapter 7, as tube power requirements are greater than those of transistors. You'll notice that this diagram does not show the tube's filament circuit. It is standard procedure to show this in the power supply circuit, as in the examples given in chapter 7.

As far as DC is concerned R1 serves only to tie the grid to B− (zero potential), and to serve as a leakage path for any electrons gathering on the grid. The current through the resistor is negligible, therefore no voltage drop takes place across it.

In the main current path through R3 and R2 you can visualize V as a variable resistor, its resistance varying with the signal voltage on the grid. When no signal is applied to the grid a steady current will flow. The values of R3 and R2 are chosen to give the proper operating voltages at the cathode and plate. A positive voltage on the cathode is provided by the voltage drop across R3, making the cathode positive with respect to the grid, which is the same as saying that the grid is negative with respect to the cathode. The value of R3 is chosen so that the voltage across it biases the grid for operation approximately in the center of the linear portion of the tube's characteristic curve (Class A operation: see Appendix).

The value of R2 depends upon the resistance of V (its plate resistance). When the resistance of V varies, the current flowing through it varies, and consequently the voltage flowing through R2. The value of R2 must be such that maximum voltage variations can be obtained across it, but if it is made too large there will be too great a voltage drop for proper operation of the tube. For a triode the maximum value of R2 should not exceed three times the plate resistance. (See further on for notes on the use of pentode tubes as voltage amplifiers.)

Similarly in Figure 2.4 electron flow is from B– through R4, Q and R3, and back to B+. However, the bias on the base is established by the voltage drop across the emitter-base junction in the transistor instead of by the voltage drop across R4, as was the case in the tube circuit. In the tube a steady plate current flows when there is no signal on the grid. In the transistor a steady emitter-base current flows under the same conditions, which in turn permits a steady (but larger) current to flow from emitter to collector. In the NPN transistor shown this will be electron current; in a PNP transistor it would be hole current. (See more about transistors in the Appendix.)

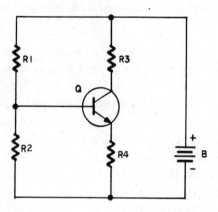

Figure 2.4 Transistor DC Circuit

The purpose of R4 is to stabilize the emitter-base current. Transistors are sensitive to temperature variations. If the temperature rises the emitter-base resistance decreases, so that the current increases. This, in turn, allows the emitter-collector current to increase. As the current increases it heats the junctions further, so their resistance decreases further, and this situation eventually results in the destruction of the transistor if steps are not taken to prevent it.

The voltage drop across the emitter-base junction will be about .2 volt for a germanium transistor, and about .7 volt for a silicon transistor. If the current flowing through R4 results in a voltage drop of 1 volt across it, the emitter will be at a potential of +1 volt. In a germanium transistor, therefore, the base voltage will be +1.2 volts. This gives the correct forward bias across the junction for Class A operation (see Appendix).

Voltage Amplifiers

If the emitter-base current increases due to increased temperature, the current through R4 increases, with a consequent rise in voltage on the emitter. But the voltage on the base is still held at +1.2 volts by the voltage divider formed by R1 and R2. With a 12-volt battery, these resistors' values are chosen so that the resistance of R1 will be nine times the resistance of R2, so nine-tenths of the battery voltage will be dropped across R1 and one-tenth across R2. One-tenth of 12 is 1.2, of course, so the voltage at the junction of R1 and R2 must be +1.2 volts. Since this voltage is also the base voltage, and remains at this figure while the emitter voltage rises (or tends to), you can see that the forward bias across the junction actually decreases. This decreases the current across the junction, and counteracts the effect of the temperature rise.

As in the case of the tube, the transistor can be visualized as a variable resistor in series with R4 and R3, its resistance varying with the signal voltage on the base. When no signal is applied a steady current will flow. However, when the resistance of Q varies, the current flowing through it varies, and consequently the current flowing through R3. The value of R3 must be such that maximum voltage variations can be obtained across it, but at the same time it must also provide for the proper operating voltage for the collector of the transistor, and, as in the case of the tube circuit, the actual value has to be a compromise between these two requirements.

This circuit uses an NPN transistor. A PNP would work just as well with the battery polarity reversed.

AC Input Subcircuit (Tube): Figure 2.5 shows the AC input subcircuit for the tube. As C1 and R1 are the means of coupling it to the previous stage, it is *RC-coupled*. C1 also prevents DC from reaching the grid of V from the previous stage. The values of C1 and R1 are chosen so that C1 presents a low reactance to signals in the audio range, and R1 has as high a value as possible without preventing it

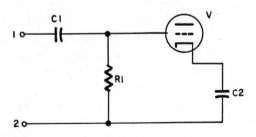

Figure 2.5 Vacuum-Tube Input Circuit

from acting as a leakage path for electrons collected by the grid of the tube.

Current cannot flow between the grid and the cathode of the tube, so the signal voltage divides between the low reactance of C1 and the high resistance of R1. Nearly all of it appears across R1, and consequently as large a portion of the available signal as possible is applied between the grid and cathode of V.

The value of C2 is such that it offers very low reactance to audio frequencies. Its purpose is to bypass the signal frequencies around R2, so it is called a *bypass capacitor*. In this way the cathode is at *signal ground* for AC, at the same time as it is *above ground* for DC.

Output AC Subcircuit (Tube): Figure 2.6 shows the output AC subcircuit. The signal voltage on the grid is alternating between positive and negative at various audio frequencies. The grid is biased to be at a negative voltage which places the tube's operating point in the middle of the linear portion of its characteristic curve. When the signal swings in a positive direction this negative bias is reduced, so the tube resistance is decreased and it conducts more current. When the signal swings in a negative direction this negative bias is increased, so the tube resistance is increased and it conducts less current.

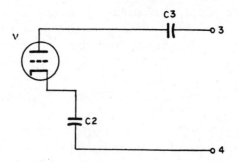

Figure 2.6   Vacuum-Tube Output Circuit

When the tube conducts more, more current flows through R2, and therefore a greater voltage is dropped across it. This makes the plate voltage less. When the tube conducts less, less current flows through R2, so the plate voltage rises. In short, a positive swing of the grid voltage results in a negative swing of the plate voltage, and vice versa. The gain of an audio voltage amplifier such as this should be close to 100, which means that the amplitude of the voltage variations at the plate is 100 times that on the grid. Remember that although they are a faithful replica, they are 180° out of phase with the grid signal voltages.

C3 can pass AC but not DC, therefore it passes the *variations* of voltage on the plate of the tube, because they are varying at audio frequencies, but it blocks the DC portion which is not varying. Since C3 is coupling the output signal into the next stage the same considerations affect the choice of its value as applied to C1.

AC Input Subcircuit (Transistor): Figure 2.7 shows the AC input subcircuit for the transistor. The remarks above concerning the tube input circuit apply equally here. However, in a transistor circuit you will find that C1 has a higher value, while R2 has a lower value. (Sometimes it is omitted altogether.) This is because the resistance between the emitter and the base is low enough to allow an appreciable current to flow, so that the transistor has a low input impedance. To avoid dropping an excessive amount of the signal voltage across C1 its value has to be such that it offers much less reactance to the signal frequency than was the case with the tube, where the grid resistor had a very high value.

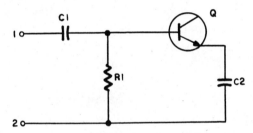

Figure 2.7 Transistor Input Circuit

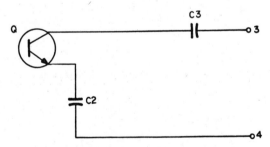

Figure 2.8 Transistor Output Circuit

Output AC Subcircuit (Transistor): In the same manner as in the tube circuit, the signal voltage on the base of the transistor in Figure 2.8 is alternating between positive and negative, resulting in variations in the voltage on the collector which reflect those on the base but are 180°

out of phase with them. R3 is the load resistor across which the output signal voltage is developed.

The gain of an audio voltage amplifier such as this should be between 50 and 100.

Circuit Variations

Although the audio voltage amplifiers just discussed are typical, you are bound to meet variations in them, though not such as to make identification difficult. For example, a pentode tube could be used in place of the triode, as in the example in Figure 2.21.

Hi-fi amplifiers usually require additional low-frequency compensation. A low-frequency limit of 100 hertz is adequate for AM radio, but a good hi-fi amplifier is expected to handle frequencies down to 20 hertz or lower, which introduces problems with the cathode-bypass resistor.

Tone controls are often found in audio voltage amplifier circuits. The reactance of a capacitor varies with its capacitance and with the frequency of the AC passing through it. This makes it possible to design circuits which can attenuate some frequencies without attenuating others.

Bass Tone Control: In Figure 2.9 the input to an audio voltage amplifier is shown with the addition of a variable resistor R1 and capacitor C2. If the slider of the variable resistor is at the upper end of R1 all the signal voltage applied across R2 will be shunted by C2. The value of C2 is chosen so that it offers little reactance to higher frequencies. Consequently the higher frequencies pass readily through C2, and their voltage across R2 is greatly reduced. The lower frequencies, meeting more reactance, are attenuated less, and the lowest not at all. As a result, the bass or low-frequency end of the audio voltage spectrum becomes predominant. As the slider of the variable

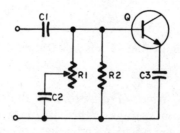

Figure 2.9   Bass Tone Control

Voltage Amplifiers

resistor is moved downwards, resistance in series with C2 increases, reducing the shunting effect across R2, so that the higher frequencies are attenuated less.

Treble Tone Control: In Figure 2.10 capacitors C2 and C3 have been added between C1 and R2, and a variable resistor R1 connected in parallel with them. The reactance of C2 is very great to lower frequencies, but becomes increasingly less to higher frequencies. When R1's slider is all the way to the right C3 is shorted out, but C2 is shunted by R1, and offers high reactance to all lower frequencies. However, if the slider is moved all the way to the left, C2 is shunted and all frequencies bypass it, and pass through C3 (which has a much higher value) without attenuation. Intermediate positions of the slider give proportionate attenuation to lower frequencies.

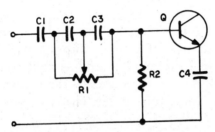

Figure 2.10   Treble Tone Control

We have analyzed this audio voltage amplifier circuit in greater detail than we shall analyze most of the other circuits in this book. This is such a basic circuit that most of its features will be found over and over again in other circuits, so that it will not be necessary to repeat the explanations given here. Instead, we shall continually refer to this section.

DIRECT-COUPLED AUDIO VOLTAGE AMPLIFIERS

In the RC-coupled audio voltage amplifiers of Figures 2.1 and 2.2, capacitors C1 and C3 isolate the DC voltages from preceding and following stages, although they transfer the AC signal with minimum attenuation. In a direct-coupled audio voltage amplifier, on the other hand, there is no DC-voltage isolation. Since this creates different problems in vacuum-tube circuits and transistor circuits, we shall deal with the vacuum-tube circuit first.

Direct-Coupled Vacuum-Tube Audio Voltage Amplifier

Figure 2.11 shows two vacuum-tube voltage amplifiers coupled directly. Other than the absence of coupling capacitors, each circuit is identical with that in Figure 2.1, except that V2 apparently has no grid resistor and its cathode resistor has been replaced by battery B2. We shall see why this is so in a moment.

Distinguishing Features

These are the same as for RC-coupled audio voltage amplifiers.

Uses

These amplifiers are especially effective at lower frequencies, because the impedance of direct coupling does not vary with frequency. They can, therefore, be used to amplify very low frequency variations in voltage. Also, because their response is almost instantaneous, they are distortion free. They are, consequently, used in

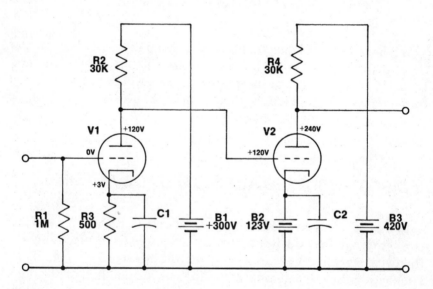

Figure 2.11 In direct-coupled vacuum-tube amplifiers the B-voltage potentials must be raised for each successive stage.

amplifiers designed to amplify DC voltages (oscilloscopes and electronic voltmeters), pulse signals and high-fidelity music.

Detailed Analysis

DC Subcircuit: For V1 the operation of the DC subcircuit is identical with that described already and illustrated in Figure 2.3. It is also the same in principle for V2, but some voltage adjustments are required to compensate for the 120 volts on V2's grid, which results from its being connected directly to V1's plate.

To get around this, the cathode and plate voltages of V2 must be increased by 120 volts. This will then restore the correct relationship between them. V2's cathode will be 3 volts positive with respect to its grid, and the plate will be 117 volts more positive than the cathode, as in V1's circuit. V2 will never know the difference! As you can see from Figure 2.11, this is done by batteries B2 and B3. However, in practice these voltages will usually be obtained from an electronic power supply with a voltage divider, as explained in chapter 7. You can also see that the presence of B2 or an equivalent voltage source makes a cathode resistor for V2 unnecessary.

AC Subcircuits: The operation of the AC subcircuits is the same as described for Figures 2.5 and 2.6 above, except for the remarks about C1 and C3. In V2's case, however, there is no actual grid resistor (R1 in Figure 2.5). This is because V1's load resistor (R2 in Figure 2.11) and the resistance of B1 have, in effect, replaced it.

Circuit Variations: Stringing identical amplifiers together is called *cascading*, therefore when stages are connected together so that the output of one amplifier becomes the input of the next, the arrangement is called a *cascade amplifier*. While there would not be any great difficulty in having a cascade amplifier with several stages as long as they were RC- coupled, it is rare to find a direct-coupled vacuum-tube cascade amplifier with more than two stages, because it is hard to achieve stable operation with more. Any small changes in the voltages of the first tube will be amplified, and thus will make it difficult to maintain the proper bias on the final tube, not to mention the increasing complications with the power supplies.

Direct-Coupled Transistor Audio Voltage Amplifiers

Cascading RC-coupled transistor audio voltage amplifiers present no problem, as you can see from Figure 2.12, which shows two of the

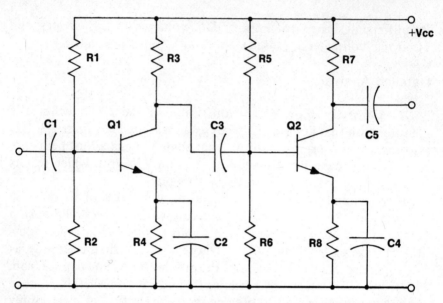

Figure 2.12 Two-Stage Transistor Amplifier Using RC Coupling. Each stage is the same as the amplifier shown in Figure 2.2. The battery symbol has been omitted. It is "understood" to be connected between the V_{CC} terminal and the low side of the circuit.

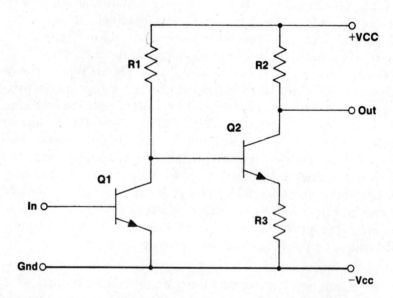

Figure 2.13 Direct Coupling Between Two n-p-n Transistor Amplifiers

Voltage Amplifiers

amplifiers of Figure 2.2 cascaded. You will notice that Q1 and Q2 have been drawn without their enclosure symbols. This is generally done nowadays in solid-state schematics, as long as it causes no confusion.

Cascading direct-coupled transistor audio voltage amplifiers are much more common than is the case with vacuum-tube amplifiers. For one thing, the power supply problem is not as complicated; and for another, the introduction of the integrated circuit (IC) has made the use of coupling capacitors impractical. To duplicate the circuit of Figure 2.12 in the form of an IC would be difficult and expensive. So, while a circuit such as this would still be used where discrete components are employed, it will not be found in a micro-electronic device.

A circuit that might be, however, is shown in Figure 2.13. In this circuit both transistors are n-p-n types, but they could be p-n-p if V_{CC} were negative instead of positive.

Distinguishing Features

Both transistors are connected as common-emitter amplifiers, and both have load resistors (R1 and R2), but Q1 does not have an emitter resistor. There are no components in the circuit to indicate that it is used for anything other than audio frequencies.

Uses

Because of the economy of parts—only five compared with 15 in Figure 2.12—it is easy to microminiaturize, and therefore is most likely to be used in a linear IC.

Detailed Analysis

DC Subcircuit: These are n-p-n transistors, so the majority carriers are electrons. Electron flow is against the arrow symbols on the emitters, so current will flow from the common return or low side of the circuit, which is connected to the battery's negative terminal, as follows: (1) via Q1 and R1 to $+V_{CC}$, connected to the battery's positive terminal; and (2) via R3, Q2 and R2, to $+V_{CC}$.

You are probably wondering how on earth the correct bias is maintained on the bases of Q1 and Q2. Actually, in Q2's case, this is not difficult to visualize, because its entire circuit is just the same as that shown in Figure 2.4, except that Q1 replaces R2. As long as the base-emitter bias on Q1 stays the same it will keep the same resistance, so to all intents and purposes it is a fixed resistor.

But if a temperature rise should cause the current through Q1 to increase, there is no emitter resistor to stabilize it, so what happens? Well, if the current through Q1 increases, the current through R1 has to increase also. This results in a greater voltage drop across R1, so lowering the collector voltage on Q1. This not only opposes the current rise in Q1 but, by lowering the bias on Q2's base, lowers the current through it also. However, this reduction of current through Q2 results in a reduction of current through R2, which causes the voltage drop across it to decrease, so Q2's output voltage rises.

This is one of the problems associated with direct-coupled circuits. Even with circuit stabilization, changes of temperature result in small changes in output, which are amplified in succeeding stages, so that an excessive change may occur in the DC level of the final output. This is called *DC drift*, and the solution to the problem is the *differential amplifier*, which we shall discuss in the next section.

AC Subcircuits: The action of these is the same as already described in connection with Figures 2.7 and 2.8, omitting references to C1 and C3. Since there is also no emitter bypass capacitor (C2 in those figures), *degeneration* is present, which increases the stability somewhat and improves the low-frequency response. Degeneration is explained in the video amplifier section, later in this chapter.

Circuit Variations: Two other versions of direct-coupled transistor amplifiers are shown in Figures 2.14 and 2.15. In Figure 2.14 we have in effect turned Q2 "upside down" by substituting a p-n-p transistor. Now the second stage is identical with the first stage, except for the reversal of polarity. Since R2 is Q2's load resistor the output is taken across it. One resistor less is required in this circuit than in that in Figure 2.13, so we have got down to only four components.

But if you thought that was economical, look at Figure 2.15. This two-transistor amplifier, known as a Darlington amplifier, has only three components! Other names for this circuit are Darlington pair, double-emitter follower and β multiplier.

This arrangement, while consisting of two discrete transistors, operates as if it were a single transistor. As you can see, the collectors are tied together, and there is really only one external base connection and one external emitter connection. Since the emitter of the first transistor is connected directly to the base of the second, the same current flows in both emitter-base circuits. Signal current entering the

Voltage Amplifiers 67

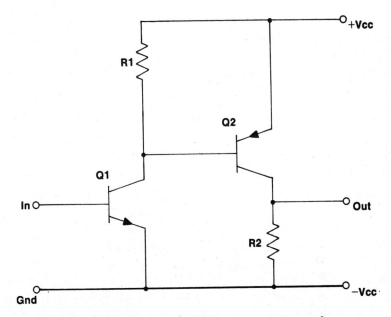

Figure 2.14   Direct Coupling Between an n-p-n and a p-n-p Transistor Amplifier

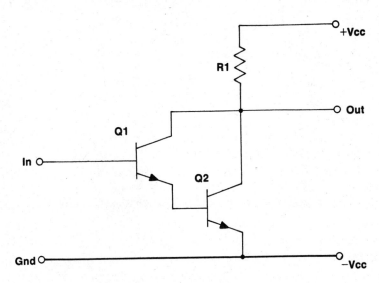

Figure 2.15   A Darlington Pair

first transistor is amplified by it, and then again by the second transistor, so their total amplification is that of Q1 multiplied by that of Q2. As mentioned previously, the gain of a transistor may be between 50 and 100, so the gain of a Darlington amplifier may be between 2500 and 10,000.

DIFFERENTIAL AMPLIFIER

A differential amplifier consists of two identical amplifiers. If identical signals are applied to the input of each amplifier, as in Figure 2.16, identical amplified signals will appear at their outputs. A voltmeter connected between the two outputs will read zero because they will be at the same potential, AC or DC as the case may be. If the two amplifiers are microelectronic circuits on a silicon chip, a temperature variation will affect each equally, since they are so small and close together. Any resulting DC drift will therefore be the same in both amplifiers, and the voltmeter will still read zero.

Since all signals that are identical cancel each other out, they are said to be rejected, and because they are common to both amplifiers this is called *common-mode rejection*. Common-mode signals must be identical with respect to both amplitude and time. Any difference between them is amplified and appears at the output, and produces a voltmeter reading.

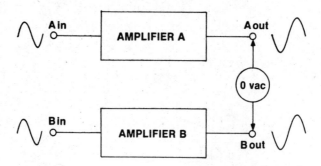

Figure 2.16 Principle of Differential Amplifier

Distinguishing Features

A differential amplifier has two inputs and two outputs. There are two vacuum tubes or transistors sharing a common cathode or emitter resistor (in microelectronic circuits this resistor is replaced by a transistor with a fixed base bias, which acts as a resistor). The two active devices are usually drawn as if one was a mirror image of the other. Direct coupling is employed between stages.

Voltage Amplifiers

Uses

Differential amplifiers are widely used in microelectronic circuits to reduce DC drift, because these circuits are direct coupled. They are also used in discrete-component circuits to reject unwanted common-mode signals. For instance, when a medical technician connects leads from an oscilloscope to two points on a patient to monitor his heart, the leads conduct the fluctuating, but different, potentials at those points to the two inputs of a differential amplifier, and the difference between them is amplified, and appears as a waveform on the oscilloscope screen. These leads also, unfortunately, pick up the 60-hertz power-line signal, and other extraneous signals in the area. In an ordinary amplifier these would be amplified and appear on the screen along with the heart waveform, in many cases obscuring it completely. But since they are common to both leads the differential amplifier rejects them.

Detailed Analysis

In Figure 2.17 (a) and (b) are shown a pair of identical amplifiers, drawn so that each is a mirror image of the other. Individually, these amplifiers are the same as the second stage of Figure 2.13.

In Figure 2.17 (c) the two amplifiers have been combined in one circuit. As a differential amplifier they now share R2, but all other components are the same.

DC Subcircuit: Direct current flow is from the low side of the circuit, which is connected to the negative terminal ($-V_{CC}$) of the power supply, via R2, Q1 and Q2, R1 and R3, to $+V_{CC}$.

AC Subcircuits: A signal applied to A_{in} causes the voltage on the base of Q1 to vary, and therefore the current flowing through Q1 varies also. This current flows through R1 as well. When the voltage on Q1's base swings in a positive direction the current through Q1 and R1 increases, resulting in a greater potential drop across R1. Since V_{CC} is the power-supply voltage and cannot change, the increased drop across R1 shows up as a reduced potential on Q1's collector. This voltage also appears at A_{out}. Conversely, when the voltage on Q1's base swings in a negative direction the voltage at A_{out} becomes more positive.

If an identical signal is at the same time applied to B_{in} the same action takes place across Q2 and R3, and since the two amplifiers are identical also, the output at B_{out} is exactly the same as the output at A_{out}. Therefore there is no output signal between these terminals.

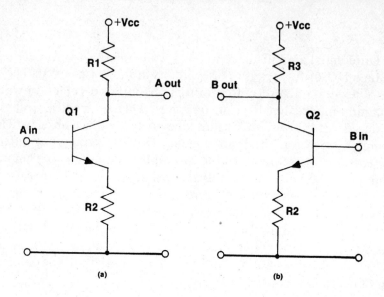

Figure 2.17 Basic Differential Amplifier (see text)

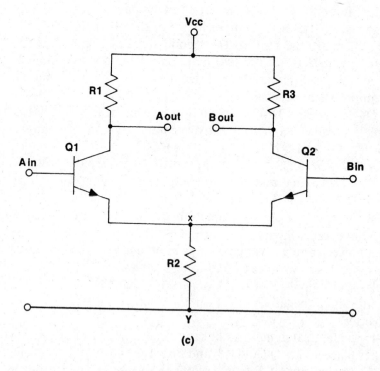

Figure 2.17 (continued)

Voltage Amplifiers

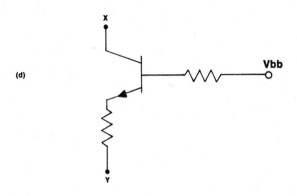

(d)

Figure 2.17 (continued)

However, it is a different story if a signal is applied *between* A_{in} and B_{in}, so that A_{in} swings positive when B_{in} swings negative, and vice versa. Now, when A_{out} swings negative B_{out} swings positive, and when A_{out} swings positive B_{out} swings negative. In other words, an amplified replica of the input signal between A_{in} and B_{in} now appears between A_{out} and B_{out}.

The resistance of R2 is always made quite large (around 10 kilohms). On the other hand, the resistances offered by R1, Q1, R3 and Q2 are quite low, so that most of the supply voltage is dropped across R2. This gives it a dominant role in establishing the level of current flowing in the circuit, so it is regarded as a *constant-current source*. As a result, variations in the resistances of Q1 and Q2 due to undesired common-mode signals have only a minor effect on the current, which is the same as saying they are not amplified very much.

However, when an input signal is applied between A_{in} and B_{in} the resistances of Q1 and Q2 change in opposite directions. Now one transistor passes more of the current flowing in R2, so less is available for the other, which is all right because it doesn't need it. In this case amplification of the desired signal is enhanced.

From this you can see that R2 is a very important component, since it cuts down the amplification of unwanted signals, but does not interfere with the amplification of legitimate ones. It is this differential treatment that gives the amplifier its name.

Circuit Variations: As you have just seen, R2 has to have a large value. In integrated circuits large-value resistors are impractical, so a transistor circuit similar to that in Figure 2.17 (d) is substituted. The fixed bias V_{BB} is chosen to allow the desired current to flow in the transistor, which to all intents is then a fixed resistor.

By connecting a fixed reference voltage to one of the inputs the differential amplifier becomes a *differential comparator*. You can read more about this type of circuit in chapter 7.

Differential amplifiers may also be designed with vacuum tubes. An example of one of these is shown in Figure 2.18. The principle of operation is exactly the same as the transistor version.

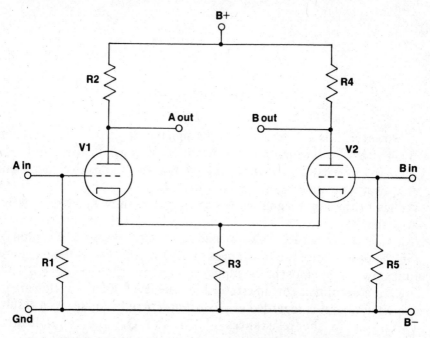

Figure 2.18   Vacuum-Tube Differential Amplifier

OPERATIONAL AMPLIFIER ("OP-AMP")

Most op-amps consist of two directly coupled differential amplifiers, like the one in Figure 2.17 (c), in cascade. That is to say, the first stage's outputs are connected to the second stage's inputs, as shown in Figure 2.19 (a). An output stage may provide additional gain but, more importantly, changes the dual input to a single output with very low impedance.

Since the circuits of modern op-amps are all very much alike, it is usually preferable to represent them by the triangular symbol of Figure 2.19 (b), which saves drawing the detailed circuit. However, this means

Voltage Amplifiers

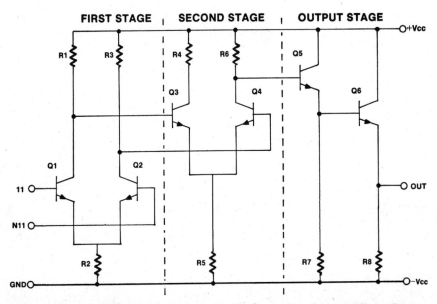

Figure 2.19(a) Simplified Schematic of Operational Amplifier. (In an integrated circuit, R2, R5 and R7 would be replaced by transistor circuits like that in Figure 2.17(d).)

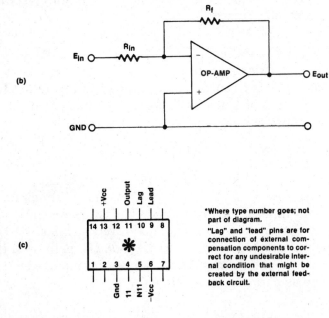

Figure 2.19(b) and (c) Feedback Circuit and Block Diagram of Operational Amplifier

Voltage Amplifiers

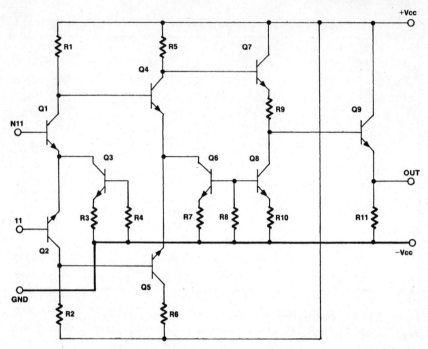

Figure 2.19(d) The Same Op-amp as in (a), but Redrawn with Transistor Constant-Current Emitter Circuits

that you have to refer to the manufacturer's data book if you want to see the internal schematic diagram.

Another symbol that is frequently used is shown in Figure 2.19 (c). This is a representation of a dual-in-line-packaged integrated circuit ("dip-chip"), appropriate if the op-amp is in that form.

Distinguishing Features

In whatever way the op-amp is diagramed, its most distinguishing feature is the two inputs and one output. The two inputs are also designated "inverting input" (II or −) and "non-inverting input" (NII or +). There is also almost always an external *feedback* path from the output to one of the inputs (in most cases, the inverting input) when the op-amp is shown as part of a larger circuit.

Uses

The op-amp was originally designed for analog computers to

Voltage Amplifiers

perform such mathematical *operations* as integration, differentiation, summation, multiplication, comparison and others, the specific function being determined by the feedback circuit. Although integrated op-amps are still used in these applications, their capabilities have been greatly expanded so that now they are the most versatile of linear IC's. They are used for active filters, biological function amplification, analog-digital conversion, peak detection, averaging, oscillator and pulse circuits, voltage-controlled oscillator circuits, function generation, and many more applications. Some of the feedback circuits used are shown in Figure 2.20. However, in this section we shall deal only with the op-amp as a DC amplifier.

Detailed Analysis

The operation of the first two stages of the circuit in Figure 2.19 (a) is the same as that of the differential amplifier in Figure 2.17 (c), so we don't need to go over it again here. These stages perform the important tasks of reducing DC drift and rejecting common-mode signals, as we have seen. They also do most of the amplifying. It is essential for op-amps to have a high voltage gain because of the use of negative feedback, explained below. For many applications this should be as high as 30,000 to 40,000 times, but in all cases several thousand times.

The third stage consists of two *emitter-follower* transistors. Emitter-follower circuits are analyzed in chapter 12, so here we'll just point out that their outputs are taken from across the emitter resistors R7 and R8. Emitter followers do not invert the signal and have a very low output impedance, which is necessary because of the feedback circuits already mentioned. Q5 is called the driver transistor because it supplies the input to the output transistor Q6.

If an input signal is applied to the inverting input it appears on the base of Q1. An amplified replica of the signal then appears on Q1's collector, and also on the base of Q3. This signal is inverted with respect to the input signal, as explained in the first part of this chapter, so when the input signal changes in a positive direction the signal on Q3's base changes in a negative direction. This causes the resistance of Q3 to increase, so current through the transistor decreases. Current through R5 remains constant, so current through Q4 increases. This results in a greater potential drop across R6, so that Q4's collector voltage decreases. In other words, the output from Q4 is inverted with respect to the input signal applied to the base of Q1. As Q5 and Q6 do not change the phase of the signal, it is still inverted at the output terminal.

Voltage Amplifiers

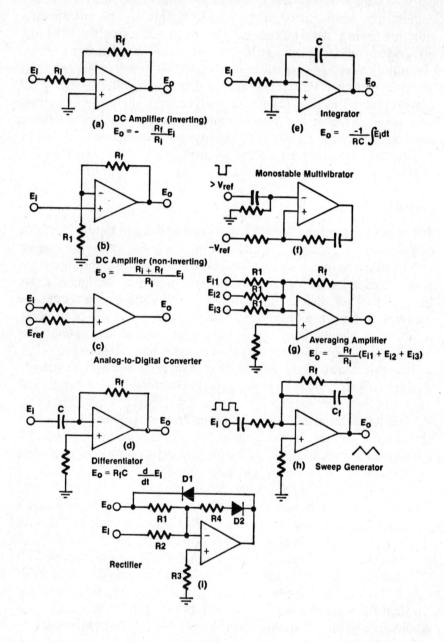

Figure 2.20 Operational Amplifier Feedback Circuits

On the other hand, if an input signal applied to the non-inverting input changes in a positive direction, Q2's output and Q4's input potentials change in a negative direction. The resistance of Q4 increases and the current through R6 decreases. This results in a lesser potential drop across R6, so Q4's collector voltage increases. As it is not inverted by Q5 and Q6, the signal at the output is of the same phase as that of the input.

In short, the output signal will be in phase with a signal applied to the non-inverting input, but of opposite phase to one applied to the inverting input.

Feedback

Because there is much more energy at the output of the op-amp than there is at its input, you can easily take a portion of its output energy and insert it into either of its inputs, as in the various examples in Figure 2.20. The op-amp then has external *feedback*.

There are two types of feedback. If the portion of the output signal voltage inserted into the input is 180 degrees out of phase with the input signal voltage, the feedback is called *negative* or *degenerative*. On the other hand, if the voltage fed back is in phase with the input signal, the feedback is called *positive* or *regenerative*. With negative feedback the voltage that is fed back opposes the signal voltage; this decreases the amplitude of the input voltage. With a smaller input voltage, of course, the output is smaller. The effect of negative feedback, then, is to reduce the amount of amplification. At first sight this may seem rather pointless, but it does have its uses, as explained in the next chapter.

Positive feedback increases the amplification, because the fed-back voltage adds to the original signal voltage, and the resulting larger input voltage causes a larger output voltage. If the energy fed back becomes large enough, a self-sustaining oscillation will be set up. This use of feedback is discussed in chapter 4.

Since the op-amp has two inputs, non-inverting and inverting, you can connect it for positive or negative feedback. In Figure 2.19 (b) the op-amp is connected for negative feedback, which is the most common arrangement. Regardless of the gain of the op-amp without feedback, which is called its *open-loop gain*, the gain with feedback will be equal to the ratio between R_f (the feedback resistor) and R_{in} (the input resistance). For example, if R_f is 50 kilohms and R_{in} 5 kilohms, the *closed-loop gain* of this circuit will be 50/5 = 10, even if the open-loop gain is in the thousands.

Circuit Variations

As in the case of the differential amplifier, the microelectronic version of the op-amp will employ transistorized emitter circuits, as in Figure 2.19 (d). Apart from this, the circuit is exactly the same as that in Figure 2.19 (a), but is drawn here with a different layout so you can see another way in which it could be diagramed. Sometimes these emitter circuits will also contain diodes to clamp the base voltage of the transistors, as explained in the section of chapter 7 dealing with voltage regulation.

The earlier analog computers used vacuum-tube circuits, so we should mention that op-amps were originally in that form. However, the state of the art has advanced so far beyond that stage that you are most unlikely to encounter any; that is, unless you are combining archeology with your electronic studies!

VIDEO AMPLIFIER

Video amplifiers are similar to audio voltage amplifiers except for additional features required to broaden the frequency response. Figure 2.21 shows a typical video amplifier using a vacuum tube. Figure 2.22 shows another with a transistor.

Distinguishing Features

Video amplifier circuits are *grounded-cathode* or *common-emitter* circuits (to obtain maximum gain), and have *load resistors* (to obtain a voltage output) in the same way as audio voltage amplifiers.

However, they are different from circuits handling only audio frequencies since they require *peaking coils* to extend their high frequency response. L1 and L2 are peaking coils, which with their associated resistors are characteristic identifiers for video amplifier circuits.

Uses

Video amplifiers are voltage amplifiers which have to amplify the demodulated television signal so it can drive the picture tube. They are, therefore, found in television sets between the picture tube and the video detector. Whereas the maximum frequency handled by audio

Voltage Amplifiers

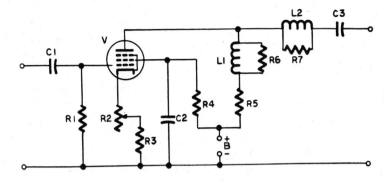

Figure 2.21   Video Amplifier with Vacuum Tube

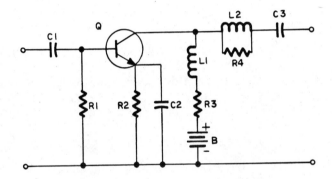

Figure 2.22   Video Amplifier with Transistor

voltage amplifiers does not usually exceed 30 kilohertz, the video frequency range extends to 4 megahertz.

Detailed Analysis

DC Subcircuits: In Figure 2.23 the electron flow is from B− through R3, R2, V and back to B+ through the parallel paths of R4 and L1−R5. The main difference between this DC circuit and the triode circuit is the additional path through R4, which drops the supply voltage to the proper value for the screen grid of the pentode.

You can visualize V as a variable resistor, its resistance varying with the signal voltage on the control grid. The resultant changes in current flowing through load resistor R5 provide the required voltage output.

A positive voltage on the tube cathode is obtained by the voltage

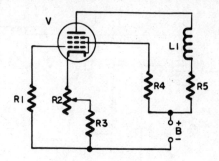

Figure 2.23 Vacuum-Tube DC Circuit

drop across R2 and R3. In the absence of a signal the steady tube plate current will give a voltage that will bias the grid for Class A operation (see Appendix).

The grid-leak resistor R1 keeps the grid at zero potential (or negative with respect to the positive potential on the cathode).

The value of load resistor R5 is considerably lower in a video amplifier than in an audio amplifier, though still several kilohms, to avoid high-frequency attenuation. The peaking coil L1 offers no impedance to DC.

Similarly, in Figure 2.24 electron flow is from B– through R2, Q, L1 and R3, and back to B+. Bias on the base is again established by the voltage drop across the emitter-base junction in the transistor. With no signal on the base, steady emitter-base current flows, which in turn permits a steady (but larger) current to flow from emitter to collector. In the NPN transistor shown this will be electron current; in a PNP it would be hole current (see Appendix).

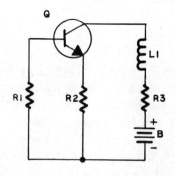

Figure 2.24 Transistor DC Circuit

Voltage Amplifiers

The purpose of R2 is to stabilize the emitter-base current.

Visualize the transistor as a variable resistor in series with R2 and R3, its resistance varying with the signal voltage on the base. When the resistance of Q varies, the current varies, including the current through R3. The value of R3 must be such that maximum voltage variations can be obtained across it, but at the same time it must provide for the proper operating voltage for the collector of the transistor, and not permit undue degeneration of high frequencies in the output circuit.

This circuit uses an NPN transistor. However, a PNP would perform just as well if the battery polarity were reversed.

AC Input Subcircuit (Tube): Figure 2.25 shows the AC input subcircuit for the tube, which is *RC-coupled* (see Figure 2.5).

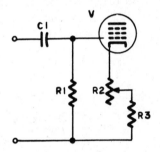

Figure 2.25   Vacuum-Tube Input Circuit

In this circuit there is no cathode-bypass capacitor as there was in Figure 2.1. The reason for this is that at low frequencies its reactance may be as high as or higher than the cathode resistor. Therefore these low frequencies would suffer *degeneration*, or *negative feedback*.

Degeneration is caused by the voltage on the cathode varying in accordance with variations in the signal voltage on the control grid. When the grid voltage swings in a positive direction the current flowing through R2 and R3 increases, so that the voltage on the cathode increases. This results in a smaller potential difference between the grid and cathode than would have been the case if a cathode-bypass capacitor had been present to ground the cathode voltage variations. When the grid voltage swings in a negative direction the current through R3 and R2 decreases, so that the voltage on the cathode decreases, again canceling out part of the potential difference between grid and cathode. Actually, the total resistance of the tube is increased, so you get a smaller output signal, but as the degeneration is the same for all frequencies the lower frequencies are not penalized.

The variable resistor R2 is a *contrast control*. It operates by varying the potential difference between the grid and cathode, thereby effectively varying the tube resistance, and ultimately the strength of the picture on the TV screen.

Degeneration has the overall result of decreasing the amplification of the circuit. For this reason a higher gain tube is required than would be the case without it. Pentodes with high gain are used in preference to triodes, and sometimes more than one stage is required. The output signal is reversed 180° with respect to the input signal.

Output AC Subcircuit (Tube): Figure 2.26 shows the output AC subcircuit for the tube. C3 couples the output signal into the next stage, and the same considerations affect its choice as apply to C1. Capacitor C2 bypasses the screen-grid resistor so that the screen grid remains at AC-ground potential for signal frequencies, to prevent degeneration. Any circuit which bypasses unwanted AC to ground is called a *decoupling circuit.*

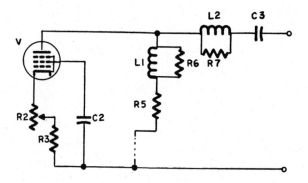

Figure 2.26 Vacuum-Tube Output Circuit

The main concern in the output circuit is to prevent attenuation of the high-frequency end of the video spectrum. The reason for this attenuation is that a certain amount of capacitance exists between the high and low sides of the circuit anyway, and the higher frequencies "leak" to ground through it. This capacitance is not due to actual capacitors, but to the natural capacitance that exists between various components and wires, and is called *distributed capacitance*. At higher frequencies the effect of distributed capacitance becomes so great that such signals bypass the load resistor R5. Consequently the output for higher frequencies is at a much lower voltage than for lower frequencies, with a great loss of fine detail in the TV picture.

Voltage Amplifiers

This can be offset to some extent by making the load resistor smaller, as mentioned before. The use of peaking coils, however, is the most important compensation.

Peaking coil L1 forms a *parallel-resonant* circuit (called *shunt peaking*) with the distributed capacitance, so as to offer increased resistance to the higher frequencies, thereby increasing the gain of the amplifier for them. The addition of resistor R6 across the coil broadens its bandpass.

L2 forms a *series-resonant* circuit (*series peaking*) with the input capacitance of the following stage, allowing for maximum transfer of signal energy at the upper end of the video spectrum.

By proper choice of inductance values, the two peaking coils can be made resonant over different frequency bands, to extend the amplifier high-frequency response linearly to cover the upper part of the frequency range required.

AC Input Subcircuit (Transistor): Figure 2.27 shows the AC input subcircuit for the transistor, which is *RC-coupled* to the previous stage. C1 will have a higher value than its opposite number in the tube circuit as mentioned in our discussion of the audio voltage amplifier.

In this circuit the emitter-bypass capacitor has been left in, so no degeneration will be experienced, except for the very low frequencies (which may be compensated for in another way, as explained in Circuit Variations). In this way, the full gain of the transistor can be realized, an important consideration since it is less than that of a pentode tube.

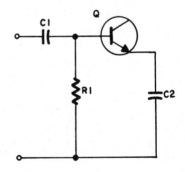

Figure 2.27 Transistor Input Circuit

Operation of the transistor is the same as for the audio voltage amplifier. The output signal is 180° out of phase with the input signal.

Output AC Subcircuit (Transistor): Figure 2.28 shows the output AC subcircuit. C3 couples the output signal to the next stage, and the

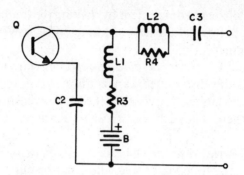

Figure 2.28 Transistor Output Circuit

same considerations apply to the choice of its value as affect the choice of C1.

As in the tube circuit the main concern in the output circuit of the transistor is to avoid attenuation of the high-frequency end of the video spectrum through losses via distributed capacitance.

Shunt-peaking coil L1 forms a parallel-resonant circuit with the distributed capacitance, so as to add to the value of the load resistor R3 for the higher frequencies for which it is resonant. The bandpass of this resonant circuit does not have to be broadened by the addition of a resistor shunting L1 as the value of R3 is already sufficiently low to load it enough. Series-peaking coil L2 forms a series-resonant circuit with the input capacitance of the following stage, allowing maximum transfer of signal energy at the upper end of the video spectrum. By proper choice of inductance values the two peaking coils are made resonant over different frequency bands to extend the high frequency response of the amplifier linearly over the higher frequencies required.

Circuit Variations

Although the video amplifiers just discussed are typical, you are bound to meet variations in them. However, you will never have any difficulty in identifying a video amplifier (a) because of its position in the overall schematic (between the video detector and the picture tube), and (b) because of the series-shunt peaking coils in the output subcircuit.

One way in which video amplifiers may vary is in the manner in which they provide for low-frequency compensation. You've seen one

Voltage Amplifiers   85

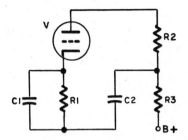

Figure 2.29 Another Type of Low-Frequency Compensation

method in the tube circuit. Another method, which is a *decoupling circuit*, is illustrated in Figure 2.29. In this circuit the value of C2 is such that at high frequencies its reactance is negligible, so that R3 is bypassed and the load resistance consists solely of R2. At low frequencies the reactance of C2 increases until R3 adds to the resistance of R2, which increases the voltage gain for these frequencies, compensating for the losses caused by C1. The values of C2 and C3 must be chosen carefully to balance out the low-frequency degeneration caused by C1 and R1.

You may be wondering how C2 bypasses R3! Although Figure 2.29 doesn't show it, the power supply is connected between B+ and the low side of C2 (refer back to Figure 2.21). The large filter capacitors in the power supply offer negligible impedance to signal voltages, so for these frequencies the low side of C2 is to all intents and purposes connected to B+.

Because of the cathode-bypass capacitor C1, the contrast control must be located elsewhere than where it was in Figure 2.21. It is often found in the picture-tube circuit instead. However, another way of doing it is shown in Figure 2.30, where bypass capacitor C1 is connected between the cathode and R2's slider, so that degeneration may be introduced as required to reduce the contrast.

Other variations may be introduced in order to provide for taking off the sound or sync signals from the video amplifier circuit, which is often done. The sound take-off may be from the plate of the tube, which will require the provision of a 4.5-megahertz parallel-resonant trap L1-C3 to block the 4.5-megahertz sound component of the television signal from getting to the picture tube.

The sync signal is usually removed from between the shunt-peaking coil L2 and the load resistor R4, as shown.

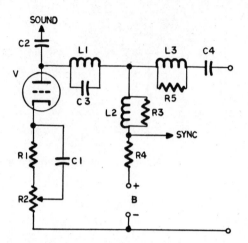

Figure 2.30   Sound and Sync Take-Off

IF AMPLIFIERS

IF (intermediate-frequency) amplifiers are voltage amplifiers which handle frequencies very much higher than audio frequencies. Figures 2.31 and 2.32 show IF amplifiers for radio. Figure 2.33 shows one for TV.

Distinguishing Features

IF amplifiers resemble other voltage amplifiers in having *grounded-cathode* or *common-emitter* circuits.

They are distinguished from other voltage amplifiers by the use of *IF transformers* to couple into and out of each stage. In Figures 2.31 through 2.33, T1 and T2 are IF transformers.

Uses

All superheterodyne receivers (this includes nearly all modern radio, TV and radar receivers) convert signals of various frequencies picked up by the antenna to a fixed lower frequency, which is then amplified by the IF amplifier. The process of conversion is dealt with in chapter 5. IF amplifiers are therefore located between the "front end" (tuner-converter section) and the audio or video portions of a receiver.

Voltage Amplifiers

As you know, audio signal information covers a much narrower range of frequencies than video. For this reason radio IF amplifiers are narrow-band amplifiers, and TV IF amplifiers are wideband.

Detailed Analysis

DC Subcircuit: In all three amplifiers illustrated the DC subcircuits are the same as those of audio voltage amplifiers, except that R1 is replaced by L2, and the current path from the plate or collector to the power source is through L3.

AC Input Subcircuit (Tube): In Figure 2.31 a double-tuned IF transformer T1 couples the input from the previous stage *(transformer coupling)*. The dashed lines surrounding T1 indicate that it is shielded. The two parallel dashed lines with the arrow, between the coils, show that the coils are *permeability tuned*, which means that within each coil is a threaded slug of powdered iron that is screwed in or out to tune the resonant circuits L1-C1 and L2-C2 to the intermediate frequency. The symbol for this transformer is different from that of an audio or power laminated iron-core transformer, which would have solid lines and no arrows (see Table I).

Some types of IF transformers are tuned with variable capacitors instead of variable inductors, but it doesn't make any difference which type it is as long as it is designed for use at the frequency of the IF amplifier in question.

The signal from the previous stage flows in L1, which is coupled inductively to L2; consequently a voltage is induced in L2. The transformer used in a vacuum-tube circuit is a *step-up* transformer (more turns on L2 than on L1), so the voltage will be higher across L2

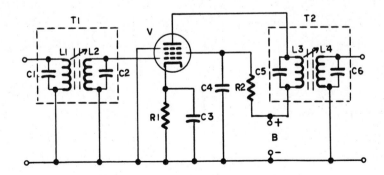

Figure 2.31   Vacuum-Tube IF Amplifier

than it was across L1. L2 and C2 form a series-resonant circuit with a resonant voltage step-up across the coil in addition to the transformer step-up, so that the signal has already received a boost before it gets to the tube. This voltage is then applied between the control grid and cathode of the pentode tube, through cathode-bypass capacitor C3.

AC Input Subcircuit (Transistor): In Figure 2.32 T1 is a *step-down* transformer, and its secondary is tapped so that the low input impedance of the transistor will not load it excessively to the detriment of its selectivity. For this reason transistor-radio IF amplification requires more than one stage as a general rule. The average is two stages for AM and three for FM. A connection through R1 allows AGC voltage (see chapter 6) to provide the proper bias on the base.

AC Input Subcircuit (Video IF Amplifier): A similar AGC connection appears in Figure 2.33. In this circuit a resistor R2 is connected so that it shunts L2. This broadens its frequency response to enable the amplifier to handle the wider bandwidth of the TV signal. See how this is exactly the opposite of the radio IF amplifiers, where steps are taken to keep the pass band as narrow as possible. You will also notice that no capacitors are connected across L1, L2, L3 and L4. At the much higher frequency of the TV IF signal (45.75 megahertz for the picture carrier) the distributed capacitance of the circuit is sufficient for the purpose.

In this circuit the cathode-bypass capacitor is omitted also, but because R3 is very small the effect on the circuit is not enough to cause degeneration.

AC Output Subcircuit: In the output circuits of all three IF amplifiers L3 forms a parallel-resonant circuit with the shunt capacitance. The output current from the tube or transistor consists of

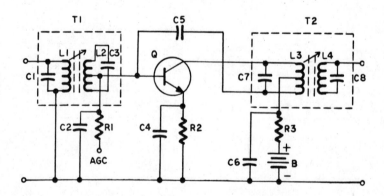

Figure 2.32 Transistor IF Amplifier

Voltage Amplifiers

a DC current with its voltage varying at an IF rate. This current really consists of two currents. One is a steady DC current of constant voltage, which forms the base for the other, which is AC fluctuating at the IF. The DC portion flows through L3 without obstruction, but the AC portion that is fluctuating at the IF encounters a high impedance because the coil has been tuned to be resonant at this frequency. Therefore this portion develops a maximum-signal voltage drop across L3. Any signals of other frequencies, however, do not meet with such an impedance, since they are not at the resonant frequency. Consequently they develop only small voltages. This is what makes this circuit selective for the chosen frequency.

The circuit in Figure 2.31 is one you'd expect to see in a vacuum-tube radio. Usually only one IF stage is required. The narrow bandwidth of an AM radio signal (10 kilohertz) or an FM radio signal (150 kilohertz) allows the use of a sharply tuned circuit, which gives maximum amplification when used with a high-gain tube.

In Figure 2.32 the transistor is connected across only part of L3. This is because the lower output resistance of the transistor acts like a resistor and would broaden the frequency response unduly if connected across the whole winding, to the detriment of good selectivity. Since the centertap on the coil is grounded for AC through C6, the two opposite ends of the winding will have signals that are 180° out of phase with each other. A connection from the lower end of the coil goes via C5 to the base of Q. This places a signal on the base which is of opposite phase to the collector signal, and serves to cancel any feedback through the transistor that might cause oscillation. (The same precaution would have to be taken with a triode tube, but is not required with a pentode because the screen grid—at AC ground potential—prevents feedback from the plate to the grid.)

In Figure 2.33 the shunting capacitor across L3 has been omitted

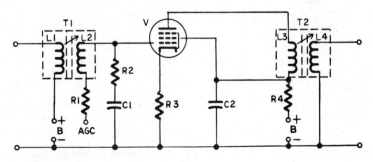

Figure 2.33   Video IF Amplifier

because the distributed capacitance is sufficient for the purpose. The loading effect of R4 in series with L3 broadens the frequency response enough for video IF amplification.

Because of the reduced amplification resulting from loading the IF transformers, video IF amplifiers require more than one stage (usually three for tubes). In a three-stage IF amplifier there will be four IF transformers. These will be *stagger-tuned*. This means that instead of all four being tuned to the same frequency, they will be tuned to different portions of the 4-megahertz bandwidth of the video signal. For example, the four frequencies might be 43.1, 43.5, 44.2 and 46.0 megahertz (they will differ slightly in different models of TV sets), which would combine in the overall picture to cover frequencies from 41.75 to 45.75, the difference of 4 megahertz required. These frequencies are usually shown on the schematic of the IF amplifier beside each IF transformer. When the picture-carrier signal is removed in the demodulator stage only the 4-megahertz band of video frequencies will remain, as explained in chapter 6.

Circuit Variations

Since the trend today is to employ integrated circuits whenever possible, you will often find your schematic looking like Figure 2.34.

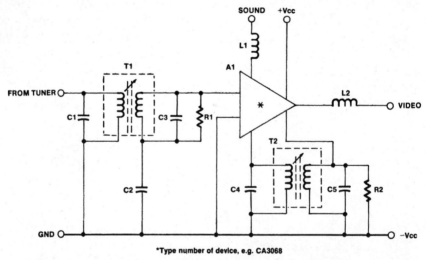

*Type number of device, e.g. CA3068

Figure 2.34   Color TV PIX-IF-System with Single IC (after RCA)

Voltage Amplifiers 91

The IC in this diagram is a dual-in-line package which contains all the active and most of the passive elements necessary for a TV receiver IF, video detector and video amplifier circuits. They are functionally similar to those in Figures 2.22, 2.33 and 6.2.

There is no way of microminiaturizing the IF transformers and their associated capacitors and resistors, so these have to be provided externally. Connector pins on the IC enable them to be connected to the proper points in the internal circuits. As you can see from Figure 2.34, they provide unmistakable distinguishing features, even though the rest of the circuit is represented only by the triangle. L1 and L2 block IF frequencies (video and sound) from reaching subsequent circuits.

RF AMPLIFIERS

RF amplifiers used in receivers are voltage amplifiers, having circuits similar to the voltage amplifiers described in the preceding portion of this chapter. In fact, RF pentode amplifiers are the same as IF pentode amplifiers, except that the input is from an antenna and tunable resonant circuit.

Most RF amplifiers in use today will be found in TV and FM tuners. For this reason Figure 2.35 shows an extremely popular type of TV circuit, and Figure 2.36 a typical solid-state FM circuit.

Distinguishing Features

In Figure 2.35 you see an RF *cascode amplifier*. This circuit uses a dual-triode tube (two tubes in one envelope). The antenna and tunable resonant circuit in the input tell you it handles RF. The first triode section is connected in a *grounded-cathode* circuit, as in other voltage amplifiers. The second is connected as a *grounded-grid* amplifier. In the output from the second tube section is a transformer (tunable to each TV channel) which couples the signals into the next stage.

In Figure 2.36 the input from the antenna and tunable resonant circuit is applied to the base of the transistor, from which you know that this circuit must also be handling RF. It is connected in a *common-emitter* circuit, with another tunable resonant circuit in the output, after which the signal is capacitor-coupled into the next stage.

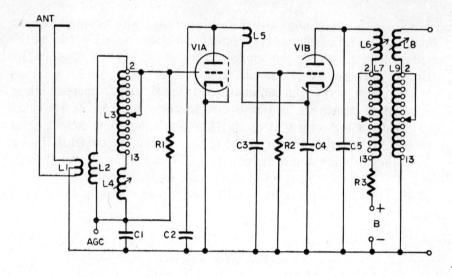

Figure 2.35 RF Cascode Amplifier

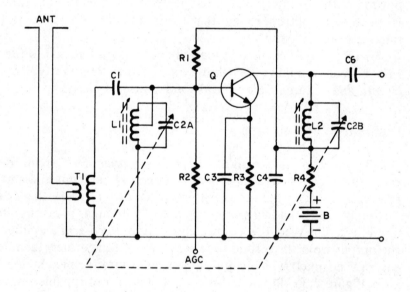

Figure 2.36 RF Transistor Amplifier

Uses

Cascode amplifiers are used extensively in TV tuners. The gain from the dual-triode is approximately the same as that from a single

Voltage Amplifiers 93

pentode, but the noise level is very much lower. However, triodes are susceptible to oscillation at very high frequencies, which has to be prevented by careful design. Pentodes were used in some older circuits before the dual-triode was developed. The two triodes in the dual-triode are shielded from each other by a screen which is grounded.

Transistors also may be used in TV tuners and modern FM radios. RF amplifiers are seldom used in AM radios, except in some automobile models. But they are used extensively in communications receivers. In any receiver the RF amplifier will be in the first stage or stages.

Detailed Analysis

DC Subcircuit (Vacuum Tube): In Figure 2.35 electron flow is from B– to the cathode of V1A, thence to the plate, and via L5 to the cathode of V1B. From V1B's plate the flow is through L6, L7 and R3 back to B+. Grid bias on V1A is supplied by AGC voltage (see chapter 6).

DC Subcircuit (Transistor): In Figure 2.36 electron flow is from B– through R3 to the emitter, through Q to the collector, thence through L2 and R4 back to B+. In a PNP transistor, with polarities reversed, the flow would be in the opposite direction. The bias across the emitter-base junction is stabilized by R1 and R3. AGC voltage is supplied via R2.

Input Subcircuit (V1A): The signals picked up by the dipole antenna are inductively coupled from L1 to L2 (Figure 2.35). This is a *balun* transformer, since it couples the *bal*anced line from the antenna to the *un*balanced amplifier circuit. (The amplifier circuit is said to be unbalanced because it has a high and low side.) The RF signals then are applied across L3 and L4, of which the upper end is connected to the grid of V1A and the lower end to the low side of the circuit via C1. L3 and L4 are the antenna section of the TV tuner, which may be a turret tuner or a rotary tuner. The sliding contact represented by the arrow selects the desired channel by shorting out more or less of L3. At the Channel 2 tap all the inductance is in the circuit; at the Channel 13 tap only L4 is left in. L4 is adjustable for touching up the overall tuning when necessary (when a tube is changed in the tuner, for example). The tuning coils L3 and L4 form a resonant circuit with the distributed capacitance of the circuit—mainly in the tube—which applies a maximum voltage between grid and cathode for the selected channel, while attenuating all the others. R1 broadens the tuning enough to give the wide pass band (about 6 megahertz in the tuner) required for the TV signal.

Output Subcircuit (V1A) and Input Subcircuit (V1B): The output of V1A is developed across V1B, which is its load resistor, as it were. This is a fairly low resistance, so there is no danger of oscillation in V1A. (Oscillation requires a strong signal in the output to give enough feedback.) Unwanted higher frequencies are blocked by L5, and decoupled by C2.

The signal is applied between the cathode and grid of V1B. Since the grid is connected to the low side of the circuit through C3, this is a *grounded-grid* amplifier circuit. The gain is not as high as in a grounded-cathode amplifier, but the grounded grid blocks feedback from the plate to the cathode, thus preventing oscillation.

Output Subcircuit (Figure 2.35): The output of V1B is applied across the RF tuning coils L6 and L7, and coupled by transformer action to L8 and L9. Tuning L7 and L9 is performed simultaneously with L3, all three moving contacts being connected to the same tuning control. R3 in series with L7 broadens the response in the same way as R1 does.

Input Subcircuit (Figure 2.36): The signals picked up by the dipole antenna are inductively coupled via the balun T1, and are then applied across the resonant circuit L1-C2A. The signal of the frequency to which the circuit is tuned is then applied between the base and emitter of Q.

Output Subcircuit (Figure 2.36): The output of Q is applied across the RF tuning coil L2, which is tuned to resonance by C2B. This variable capacitor is ganged with C2A (as indicated by the dashed lines) so that both input and output circuits are simultaneously tuned to the same frequency. The output signal is then coupled into the next stage through C6.

Voltage Amplifiers

TROUBLESHOOTING TABLE FOR VOLTAGE AMPLIFIERS	SYMPTOM							
	No output	Weak output	Distorted output	Intermittent	Hum (tubes only)	Motorboating (tubes only)	Interference	Wrong station
Vacuum Tube	X	X	X	X	X			
Tube socket	X			X				
Transistor	X	X	X	X				
Input coupling capacitor	X		X	X				
Grid resistor	X		X	X			X	
Cathode resistor	X			X				
Bypass capacitor	X		X	X	X			
Load resistor	X		X	X				
Output coupling capacitor	X	X	X	X				
Battery	X	X	X					
Power supply	X	X	X		X			
IF transformer (detuned)		X						
IF transformer (defective)	X			X				
Tuner	X	X		X			X	X
Antenna		X		X				

3
POWER AMPLIFIERS

Power amplifiers are *current* amplifiers. The vacuum tube, transistor or other device used in a power amplifier circuit must be capable of handling much greater current than a voltage amplifier. Consequently, power tubes and transistors are generally larger physically and must dissipate more heat. Distortion is also a problem requiring special treatment, but efficiency is higher.

SINGLE-ENDED AUDIO POWER AMPLIFIERS

Figure 3.1 shows a single-ended audio power amplifier using a pentode tube. Figure 3.2 shows a similar amplifier using a transistor.

Distinguishing Features

Single-ended audio power-amplifier circuits are *grounded-cathode* or *common-emitter* circuits, to obtain maximum gain. They differ from audio voltage amplifiers in not having load resistors. Instead, an *output transformer* is used to transfer *power* from the active element to the driven device (in this case a loudspeaker), which is the real load.

Power Amplifiers

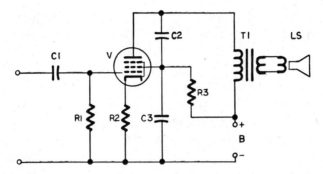

Figure 3.1 Single-Ended Audio Power Amplifier Using Pentode

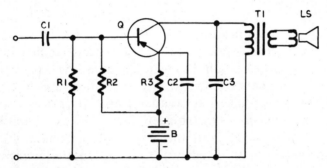

Figure 3.2 Single-Ended Audio Power Amplifier Using Transistor

Detailed Analysis

DC Subcircuit (Vacuum Tube): In Figure 3.1 electron flow is from B−, through R2, V, and back to B+ via the parallel paths through the primary winding of T1 and the screen-grid resistor R3. Grid bias for Class A operation (see Appendix) is established by the voltage drop across R2, and R1 serves as a grid leak. Variations in the resistance of V are caused by variations in the signal voltage applied between the grid and cathode, resulting in variations in the current flowing in R2 and the primary winding of T1. The value of R2 must be comparatively low because of the heavier current it has to pass.

DC Subcircuit (Transistor): In Figure 3.2 electron flow is from B−, through the primary winding of T1, Q and R3, and back to B+. Bias for

Class A operation is established by hole current (this is a PNP transistor), and stabilized by R2 and R3. Variations in the resistance of Q are caused by variations in the signal voltage applied between the base and emitter, resulting in variations in the current flowing in the primary of T1. The value of R3 must be comparatively low because of the heavier current it has to pass.

AC Input Subcircuits: Both vacuum-tube and transistor amplifier inputs are RC-coupled from the previous stage in the same way as audio voltage amplifiers. In the tube circuit the cathode-bypass capacitor has been omitted to obtain degeneration, as explained in the video amplifier analysis in chapter 2. In the transistor circuit the bypass capacitor has been retained to keep the emitter at ground potential as far as the signal is concerned, and because degeneration would still further reduce the amplification of the transistor, which is not so great as for a pentode. Since the heavier current through R3 would still cause considerable voltage variations at low frequencies, the capacitance of C2 must be large. (An electrolytic capacitor is often used.)

AC Output Subcircuits: The output transformer used in each amplifier is an impedance-matching device. The real load is the loudspeaker. Loudspeaker impedances are very low, from 4 to 16 ohms as a rule, whereas the plate impedance of a tube or the output resistance of a transistor is likely to be 1000 times greater. To match these different impedances (which is necessary if all the power is to be transferred) the output transformer must have many more turns on its primary winding than it has on its secondary. The *turns ratio* is equal to the square root of the impedance ratio. If the latter were 1000:1, then the turns ratio would have to be approximately 32:1, in theory, although it is usually made somewhat higher in practice to cut down on distortion.

The current flowing in the primary winding of the transformer consists of DC current flowing from the tube to the power source, varying in strength because of variations in the resistance of the tube. You can look at this current as consisting of two parts: a base of unvarying DC, and a superstructure of varying AC. The unvarying DC meets little resistance as it flows through the transformer winding, and becauses it causes no flux variations in the transformer core it has no effect on the secondary winding. The superimposed AC, however, is varying all the time, and the flux variations it induces in the core induce similar current variations in the secondary. Consequently the AC component is transferred to the secondary, but not the DC component.

Power Amplifiers

This AC component is the current that drives the voice coil of the loudspeaker.

In Figure 3.1, C2 is a plate-bypass capacitor and in Figure 3.2 C3 is a collector-bypass capacitor. Their value is chosen so that it will provide a path to ground for high-frequency signals which might cause high-frequency oscillation.

PUSH-PULL AUDIO POWER AMPLIFIERS

Single-ended audio power amplifiers are limited to low-power applications. Where more power is required push-pull power amplifiers are used. Figure 3.3 shows a typical push-pull audio power amplifier using tubes, and Figure 3.4 shows one using transistors.

Distinguishing Features

Two tubes or transistors sharing one output transformer, in a circuit consisting otherwise of resistors and capacitors only, indicates a

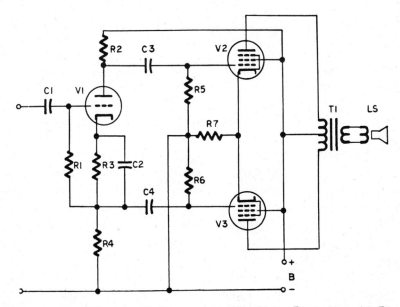

Figure 3.3 Push-Pull Audio Power Amplifier Using Pentodes, with Phase Inverter

Power Amplifiers

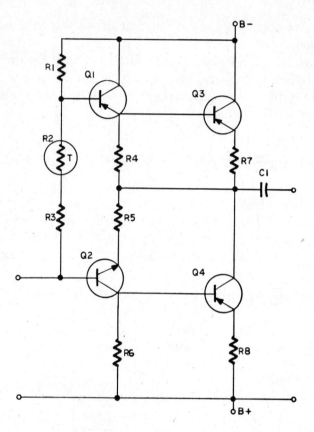

Figure 3.4 Push-Pull Audio Power Amplifier Using Transistors, with Phase Inverter

push-pull audio output amplifier. However, many transistor circuits now eliminate the output transformer by using low-collector-resistance transistors, but they are still easily recognizable as audio power amplifiers because of the doubling-up of transistors and because they are the final stage in the amplifier before the loudspeaker or other output.

Uses

Push-pull audio power amplifiers produce more power with less distortion than do single-ended amplifiers. This power can be further enhanced by doubling the active elements (two push-pull pairs in parallel), or by using more powerful tubes or transistors. Some power amplifiers are used to drive vibration test equipment, in which considerable loads are shaken for long periods to test the reliability of their construction. Another advantage of push-pull circuits is that they cancel out much of the harmonic distortion that is generated in pentode and beam-power tubes. Because of these factors push-pull output stages are almost invariably used in hi-fi amplifiers.

Detailed Analysis

In a push-pull circuit the input signal has to be split to feed the two output active elements. After splitting, the two signals must be of opposite phase, so that when one is swinging positive the other is swinging negative. In this way the outputs of the two tubes or transistors, applied at opposite ends of the output-transformer primary winding or other load, are of opposite polarity. (If they were of the same polarity they'd cancel each other out!) Consequently, in all push-pull circuits there must be some means of phase-splitting to obtain phase inversion for one of the output tubes. We shall discuss various methods of doing this in the following section.

DC Subcircuit (Tube): In Figure 3.3 the DC path for the two output tubes V2 and V3 is from B− to R7, which is the common cathode resistor for both. The current then divides and passes through each tube to the opposite ends of the primary winding of T1, joins again at the centertap, and returns to B+. The screen-grid supply follows the same route, except that it bypasses the output transformer. The value of the cathode resistor will be one that gives Class AB operation (see Appendix) and some degeneration to reduce distortion.

The DC path for V1 is from B− through R4, R3, V1 and R2, then back to B+.

DC Subcircuit (Transistor): The DC path in Figure 3.4 runs from B− at the top of the diagram to B+ at the bottom. There are three paths. The main path through the output transistors Q3 and Q4 also passes through R7 and R8. These two resistors are low value (often less than one ohm) and high wattage. They are current-limiting resistors, to protect the circuit from the effects of a sudden surge of current from B− to B+.

The second path is via Q1, R4, R5, Q2 and R6. Both these paths are designed so that in a no-signal-applied condition the voltage at the point between R4 and R5 will be half the B− voltage. (In some circuits a variable resistor is provided to adjust it, if necessary.)

The third path is via R1, R2 and R3, and the base-emitter junctions of Q1 and Q2, and is for the purpose of stabilizing the bias on these transistors. The thermistor R2 is mounted on the heat sink of the output transistors so that as the temperature rises the resistance of R2 goes down (this being the virtue of thermistors). This compensates for increases in resistance with temperature in other circuit elements.

AC Input Subcircuit (Tube): The signal is RC-coupled from the previous stage to the grid of V1. This tube is operated as an ordinary Class A audio voltage amplifier. A negative swing of the signal voltage on the grid results in a positive swing of the plate voltage, and vice-versa. However, current flowing through R4, which is not bypassed, varies directly with the grid voltage. As a result, the signals coupled through C3 and C4 to the grids of V2 and V3 are of opposite phase. Proper choice of values for R2, R3 and R4 ensures that their amplitudes are equal.

AC Output Subcircuit (Tube): With identical signals of opposite phase applied to their grids, V2 and V3 produce output signals which are not exactly symmetrical (because they are being operated as Class AB amplifiers), but which combine in the output transformer to give a true replica of the input signal. T1 couples it inductively to the load.

AC Input Subcircuit (Transistor): Q1 is a PNP transistor, and Q2 is NPN. Their characteristics are identical, and complement each other in what is called *complementary symmetry*. Each base-emitter junction has a small forward bias when no signal is applied, and under this condition a small current flows from B− through both transistors to B+.

When a signal is applied to the input it changes the bias on each transistor, causing the currents through them to change. When the bias changes in a positive direction it opposes the forward bias on Q1, and increases it on Q2. This results in an increase in the resistance of Q1 and a decrease in that of Q2. This makes the voltage on the emitter of Q1 become less negative, so that it in turn opposes the forward bias on the emitter-base junction of Q3. At the same time the collector voltage on Q2 becomes more negative, which increases the forward bias on the emitter-base junction of Q4. When the signal voltage swings the bias in a negative direction, the opposite happens in each case.

The result is that Q3 conducts more on negative-going swings of the input signal voltage, while Q4 conducts more on positive-going swings.

Output Subcircuit (Transistor): The output circuit in Figure 3.4 is extremely simple. As Q3 and Q4 alternate between conducting more and conducting less, the voltage on the collector of Q4 swings back and forth from less negative to more negative than the resting voltage.

Power Amplifiers

These are audio-frequency variations, of course, and they are coupled through C1 (an extremely high-value capacitor of probably 1000 microfarads) to a loudspeaker connected across the output terminals.

INVERSE FEEDBACK

Inverse or negative feedback is the method used to reduce distortion in an audio power amplifier. There are two kinds. In Figure 3.5 a signal A is applied to the input of the amplifier. The output signal B is an

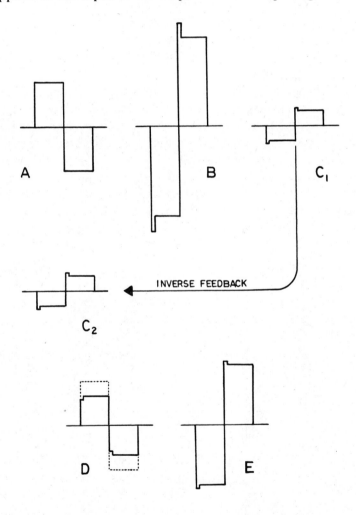

Figure 3.5   Using Inverse Feedback to Correct Amplitude Distortion

enlarged inverted replica of A, but a spurious spike has appeared on the leading edge of the square wave. This is *amplitude distortion*: the amplitude of the output signal is not a faithful (enlarged) reproduction of the input signal. By feeding back to the input a portion C_1 of the output signal, the distortion can be reduced. In the input the feedback signal C_2 combines with the input signal A and modifies it. Being of opposite phase, it is subtracted from A to produce a new signal D. The portions enclosed by the dotted lines have been canceled out, and the modified input signal now has notches where the feedback signal had spikes. When the modified signal is amplified, the notched portions receive the same excessive amplification that gave the high spikes on B, but now the spikes are much reduced, as in E.

Amplitude distortion does not necessarily take the form of spikes. It is any deformation of the shape of the signal as it passes through the amplifier.

The other form of distortion is illustrated in Figure 3.6. A good amplifier should amplify all frequencies equally. Curve A shows that the higher frequencies are being amplified more than the lower. By supplying inverse feedback from the output to the input in the same way as in reducing amplitude distortion, we can flatten the response. In Curve B we partially reduce the distortion by moderate feedback; in Curve C we eliminate it almost entirely by using more feedback.

You can see that inverse feedback also has the disadvantage of reducing the gain of the stage. In hi-fi equipment a great deal of

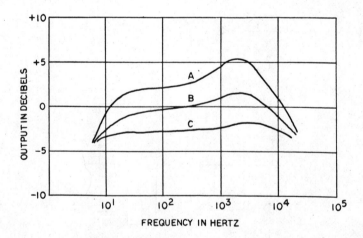

Figure 3.6 Using Inverse Feedback to Correct Frequency Distortion

feedback is used to reduce distortion. Consequently the audio signal must be given increased amplification to compensate for the loss. Fortunately the distortion reduction remains at a satisfactory level.

There are two types of inverse feedback. You are already familiar with *inverse current feedback* under the name of degeneration, obtained by omitting the cathode-bypass capacitor. It is called current feedback because the feedback signal is obtained from the current across the cathode resistor, which is varying with the signal voltage. As you saw in chapter 2, inverse current feedback, or degeneration, is often used to improve frequency response.

The second method is called *inverse voltage feedback,* and is obtained by feeding part of the output voltage back to the input, using a circuit such as that in Figure 3.7, where the feedback is via R1 to the grid of the tube from the junction of R3 and R4. Capacitor C3 blocks the DC voltage only, so R3 and R4 are connected between the plate output and ground, or the low side of the circuit. The amplitude of the voltage appearing at their junction will depend upon their relative values. For example, if the required feedback voltage is 25 percent of the output signal, R3 will be three times the value of R4.

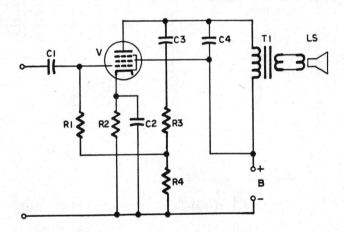

Figure 3.7 Inverse Voltage Feedback

Inverse voltage feedback can also be supplied to the previous stage, or even to several stages. In hi-fi equipment, as we've already said, a great deal of feedback is used, and it reduces the gain of each stage considerably, but as the paramount requirement is fidelity the extra expense of additional stages has to be accepted.

Circuit Variations

Another method of obtaining phase inversion for a vacuum-tube push-pull power amplifier is shown in Figure 3.8. Each half of the dual-triode tube V1 is connected as a voltage amplifier. The input signal is applied between the grid and cathode of V1A, and the output signal is then fed to the grid of V2. This signal is reversed in phase with respect to the input signal. A portion of it is fed back to the grid of V1B, where it is amplified as the original input signal was in V1A, and the output then goes to the grid of V3. This output is also reversed in phase with respect to the input to V1B, but as the input signal to V1B was a portion of the input to V2, you can see that V3's input will now be of opposite phase to that of V2, which is what we want.

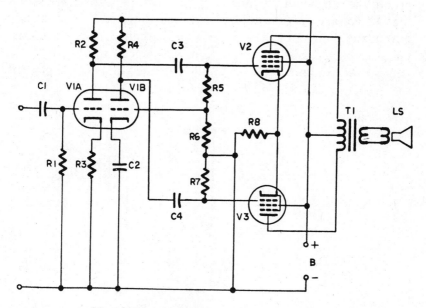

Figure 3.8   Dual-Triode Phase Inversion

The amplitude of the input signal to V1B must be equal to the amplitude of the input signal to V1A. This is obtained by choosing the values of R5 and R6 so that their ratio gives the required reduction. For example, if V1A amplifies the input signal 60 times, the ratio of R5 to R6 must be 59 to 1 to get an input signal of the right amplitude to apply to V1B, so that when it has amplified it 60 times, the signal fed to V3 will be the same amplitude as that supplied to V2.

Unfortunately, the two halves of the dual-triode tube may not

Power Amplifiers

amplify equally, especially as the tube ages. For this reason this method of obtaining phase inversion is inferior to that using a single tube. However, since each half of the dual-triode is operating as a voltage amplifier it saves having a preceding amplifier stage. The choice therefore boils down to a mattter of dollars and cents.

Phase inversion may also be obtained by using a transformer, as in Figure 3.9, where an interstage transformer T1 is shown with its secondary divided so that the upper half drives Q1 and the lower Q2. The centertap is connected via R2 to the two emitters. When one end of the secondary of T1 swings in a positive direction the other end swings in a negative direction, and vice-versa. Consequently, the signals applied to Q1 and Q2 will be of opposite phase.

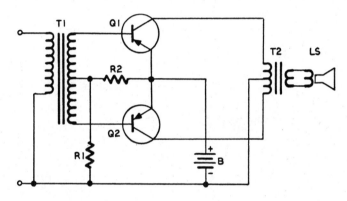

Figure 3.9   Transformer Phase Inversion

SINGLE-ENDED RADIO-FREQUENCY POWER AMPLIFIERS

Figure 3.10 shows a single-ended RF power amplifier using a triode tube.

Distinguishing Features

Single-ended RF power amplifiers are *grounded-cathode* circuits to obtain maximum gain. Input and output subcircuits are *tuned resonant circuits*: the output resonant circuit is usually a *tank circuit* feeding an *antenna*.

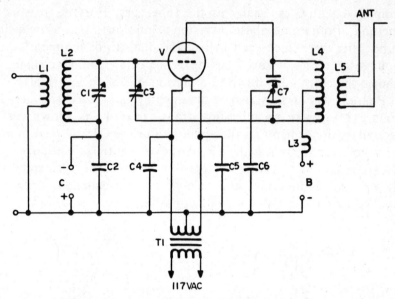

Figure 3.10 Single-Ended RF Power Amplifier

Tubes used in RF power amplifiers often require *neutralization* (see explanation in detailed analysis of the circuit). This is invariably the case with triode tubes. An RF power amplifer is operated with a much higher negative bias on the grid than other power amplifiers, requiring in most cases an *external source of grid bias voltage*.

Microwave power amplifiers are the easiest of all to identify. Their use of magnetrons, klystrons, traveling-wave tubes and waveguides is a dead giveaway.

Uses

Single-ended RF power amplifiers are operated as Class C amplifiers (see Appendix), and are capable of delivering moderate power (depending on the tube) to an antenna, and are used in many types of transmitters of low or moderate power.

Detailed Analysis

DC Subcircuit: In Figure 3.10 electron flow is from B− through both halves of T1's secondary to the filament of V, then from the plate via

Power Amplifiers

the upper half of L4 and L3 to B+. Grid bias voltage is supplied by a separate source C as shown, via L2.

Filament Subcircuit: The plate voltage used in transmitter tubes is much higher than that used in other types, and tubes of special design are required. Dissipation of heat necessitates tubes equipped with fins or even water cooling. Oxide-coated cathodes, indirectly heated, have to be replaced with heavy tungsten filaments which can stand up to the high voltage. To heat such a filament to incandescence a filament transformer is needed to reduce the line voltage to the filament voltage used by the tube. In Figure 3.10 T1 is the filament transformer. The centertap on the secondary is connected to the low side of the circuit, to provide a path for the plate voltage, and the two ends of the secondary are connected to the filament. Both ends of the filament are also connected to the low side of the circuit through C4 and C5, which provide a path for the signal voltage but not for the filament voltage.

AC Input Subcircuit: Coil L1 is inductively coupled to L2. L2 and C1 form a series-resonant circuit, connected between the grid and filament of V via C2, C4 and C5. Maximum input signal voltage is developed when L2 and C1 are tuned to the signal frequency.

AC Output Subcircuit: C7 and L4 form a parallel-resonant circuit. C7 is a split-stator variable capacitor. The rotor is connected to the low side of the circuit through C6, and the two stators are connected to the opposite ends of L4. The powerful current pulses from the tube become sine waves in the resonant circuit C7-L4 (as explained in chapter 4), and are coupled inductively to L5, and thence to the antenna. L3 is an RF choke to keep the RF signal out of the power supply.

Neutralization Subcircuit: The lower end of L4 is connected via C3 to the grid of V. Since V is a triode, energy from the plate is fed back internally to the grid, and would cause V to oscillate as it is in phase with the grid signal. However, the signal fed back via C3 is of opposite phase (being from the opposite end of L4 to the plate), and by adjusting C3 can be made exactly equal to the positive feedback, which is therefore neutralized. C3 is consequently a *neutralizing capacitor*.

Circuit Variations

Circuits similar to Figure 3.10 will be seen also with tubes with indirectly heated cathodes (in which no filament subcircuit is required), or using tetrodes or pentodes (in which neutralization may not always be necessary).

PUSH-PULL RF POWER AMPLIFIER

Figure 3.11 shows a push-pull RF power amplifier. As in the audio power amplifiers discussed earlier in this chapter, they also require some type of phase inversion.

Distinguishing Features

Push-pull RF power amplifiers are quite similar to audio push-pull power amplifiers, except for the *resonant circuits* used in the input and output, and provision for *neutralization* and an *external grid-bias voltage source*.

Uses

Push-pull circuits not only increase power output but also minimize distortion. This is especially true for triodes. Successive push-pull RF power amplifier stages are employed in transmitting systems between the oscillator and the antenna to develop the output power required. They are operated as Class C amplifiers (see Appendix).

Detailed Analysis

DC Subcircuit: In Figure 3.11 electron flow is from B− to the cathodes of V1 and V2, thence from each plate to the opposite ends of L4, returning to B+ via the centertap and L3.

Filament Subcircuit: T1 is the filament transformer, with secondary centertap connected to the low side of the circuit. C5 and C6 bypass the signal voltage around the transformer.

AC Input Subcircuit: Coil L1 is inductively coupled to L2. L2 and C1 form a series-resonant circuit. C1 is a split-stator variable capacitor, with the two stators connected across the two halves of L2. The rotor of C1 is connected via C2, C5 and C6 to the tube filaments. When the signal induced in L2 is such that its upper end is positive, the lower end will be negative, and vice-versa. In this way signals of opposite phase are applied between each grid and filament of V1 and V2.

AC Output Subcircuit: C8 and L4 form a parallel-resonant circuit. C8 is a split-stator variable capacitor. The rotor is connected to the low

Power Amplifiers

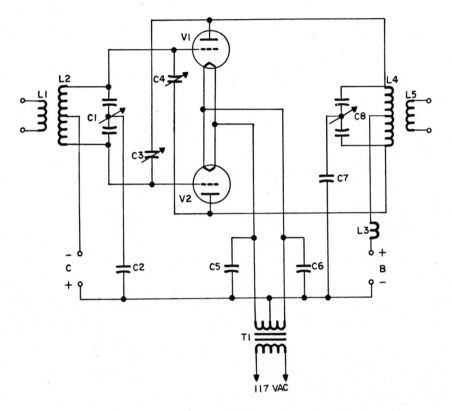

Figure 3.11 Push-Pull RF Power Amplifier

side of the circuit through C7, and the two stators are connected to the opposite ends of L4. The powerful current pulses from the tubes become sine waves in the resonant circuit C8-L4, and are coupled inductively to L5, and thence to the antenna. L3 is an RF choke to keep the RF signal out of the power supply.

Neutralization Subcircuit: The plate of V1 is connected to the grid of V2 via C3, and the plate of V2 is connected to the grid of V1 via C4. This is called *cross-neutralization*. It provides signals of opposite phase on each grid to neutralize the tube's internal plate-grid feedback. C3 and C4 can be adjusted for exact neutralization.

Circuit Variations

Circuits similar to Figure 3.11 will be seen also with tubes with indirectly heated cathodes (in which no filament subcircuit is required), or using tetrodes or pentodes (in which neutralization may not always be necessary).

LINEAR RF AMPLIFIERS

Figure 3.12 shows a single-ended linear RF amplifier. In this example a pentode is used, but triodes and tetrodes may be used as in other RF power amplifiers.

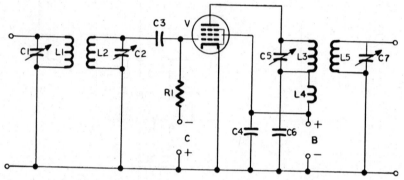

Figure 3.12   Single-Ended Linear RF Amplifier

Distinguishing Features

Linear RF power amplifiers are a special form of RF power amplifier, and are therefore similar in their circuitry. You can tell a linear RF power amplifier, however, by its position in the overall schematic. *Any RF power amplifier after the point where the carrier is modulated* (see chapter 5) *must be a linear amplifier.*

Uses

As explained above, linear amplifiers, operated Class B (see Appendix), are used to bring the power of the modulated carrier up to the level required for transmission, where this has not already been done. You can't do this with a Class C amplifier once the carrier has been modulated, because some of the modulation will be lost.

Power Amplifiers

Detailed Analysis

DC Subcircuit: In Figure 3.12 electron flow is from B– to the cathode of V. From V one path goes from the plate via L3 and L4 (an RF choke to keep the RF signal out of the power supply) back to B+. The other path goes from the screen grid directly to B+.

AC Input Subcircuit: This circuit is similar to that of Figure 3.10.

AC Output Subcircuit: This circuit is similar to Figure 3.10.

Neutralization: Neutralization will be required for all triodes and many pentodes. In Figure 3.12, however, we have shown a circuit without neutralization for comparison with Figure 3.10.

Circuit Variations

Push-pull RF linear power amplifier circuits are similar to other RF push-pull power amplifier circuits, and are used in the same way to obtain greater power output and efficiency with less distortion than with single-ended circuits.

MAGNETIC AMPLIFIERS

In magnetic amplifiers you see a power amplifier that does not use a vacuum tube or transistor to introduce power into the circuit. Instead, a special type of transformer is used. Figure 3.13 shows a half-wave magnetic amplifier, Figure 3.14 a full-wave one.

Distinguishing Features

Magnetic amplifiers consist of *transformers* with three or four windings. One (or sometimes two) of these windings is the *control winding*, which is supplied with *direct current*. This is quite a distinguishing feature, for a transformer using DC is practically never met with in any other connection. (Do not let the practice of routing the plate supply to a tube through a power output transformer, which we've just been looking at in other types of power amplifiers, confuse you, because both DC and AC are flowing in the primary of these.)

A further distinguishing feature, of course, is that there are no tubes or transistors associated with the transformer in the circuit.

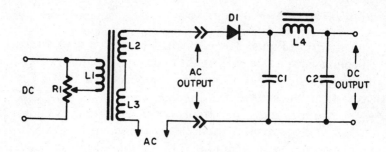

Figure 3.13 Half-Wave Magnetic Amplifier

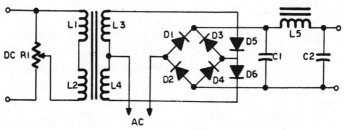

Figure 3.14 Full-Wave Magnetic Amplifier

Uses

Before the advent of semiconductor devices (see chapter 7), magnetic amplifers were the principal means whereby a large amount of power could be controlled by a quite small amount of DC. They were not used for amplifying audio or other signals, but for controlling power used in industrial applications. For example, a magnetic amplifier might be used to control the AC current supplied to a heater, maintaining a constant temperature by means of a DC current derived from a temperature sensor. Magnetic amplifiers are very rugged, and many are still in use.

Detailed Analysis

In Figure 3.13 alternating current from a 117-volt source flows through L2 and L3 and whatever load is connected across the output terminals (AC output). If DC is required, the rectifying circuit to the right of the AC output terminals will also be required. In this case the

Power Amplifiers

current returns via L4 and semiconductor diode D1. D1 will only allow current to flow one way, in the direction opposite to the arrow, so only the negative-going halves of the AC sine wave can pass. These are smoothed into DC by the π-filter consisting of C1-L4-C2 (see chapter 7).

As these pulses travel through L2 and L3 they encounter *inductive reactance*. This reactance varies inversely with the strength of the current flowing in L1. As the DC current in L1 is increased (by adjusting R1) flux is increased in the transformer core. The core is designed to have the ability to accommodate just so much flux, so that as the DC currrent in L1 increases, more and more of the *permeability* (capability of accommodating flux) is used up, and what is left decreases. This progressively reduces the inductance of the transformer, which in turn reduces the reactance it is able to oppose to the current flowing in L2 and L3. Consequently, as we increase the DC current in L1 the AC current increases in L2 and L3. Maximum current flows when inductance is reduced to zero, at which point we say the core is *saturated*.

In DC applications the circuit in Figure 3.13 is less efficient than that in Figure 3.14. This is because it can only use half of the available AC, since D1 blocks all the alternate half-cycles. In Figure 3.14 this disadvantage is overcome by using both half-cycles, positive and negative, of the AC input.

This is done by means of the four-diode bridge consisting of D1, D2, D3 and D4. When the right-hand side of the AC input goes negative, electrons flow through D2 to the low side of the output. They return via L5, D3, D5 and L3 to the left (positive) side of the AC input.

When the left-hand side of the AC input goes negative electrons flow through L4, D6 and D4 to the low side of the output. They return via L5 and D1 to the right (now positive) side of the AC input.

In this way both half-cycles of the AC sine wave are used, which is twice as efficient as in the half-wave rectifier. However, there is no loss of efficiency in the circuit of Figure 3.13 when used for an AC output, of course.

TROUBLESHOOTING TABLE FOR POWER AMPLIFIERS	SYMPTOM							
*As heard in a speaker. For non-audio power amplifier read equivalent symptom as seen on oscilloscope.	Output absent or weak	Output intermittent	Output distorted	Oscillation Howling*	Hum*	Buzz*	Hiss*	Rasping or Crackling*
Vacuum Tube	X	X	X	X	X	X	X	
Transistor	X	X						
Coupling Capacitor	X	X						
Bypass or Decoupling Capacitor			X	X	X			
Defective Resistor	X	X					X	X
Feedback Capacitor	X	X	X				X	X
Broken Wire or Defective Insulation	X	X						
Defective Control								X
Misadjusted Control	X							
Dirty Contacts Anywhere	X	X						X
Incorrect Operating Voltages			X				X	
Defective Power Supply					X	X		X
Defective or Unsuitable Speaker	X		X					X
Neutralization Misadjusted				X				

4

OSCILLATORS

Since oscillators generate their own signals they do not require an external input. This is the most obvious difference between oscillators and the amplifiers you read about in chapters 2 and 3. (However, you'll see that some oscillators have to be synchronized with an external signal, so this "rule" is not always true.)

There are two types of oscillators: *L-C oscillators*, in which frequency is determined by an inductance-capacitance network; and *R-C oscillators*, in which frequency is determined by a resistance-capacitance network.

In both types the active element is a vacuum tube or transistor acting as an automatic switch that chops the DC supply current into a series of pulses. In most R-C oscillators the output will be in the form of pulses of various shapes. In an L-C oscillator the resonant circuit transforms the chopped DC into sine waves, as you will see in the first circuit we discuss: the *Hartley oscillator.*

L-C OSCILLATORS

HARTLEY OSCILLATOR

The Hartley oscillator is one of the most important and widely used oscillator circuits. Figures 4.1 and 4.2 are two examples of this circuit.

Distinguishing Features

The Hartley oscillator employs a vacuum tube or transistor connected in a *grounded-cathode* or *common-emitter* circuit.

There is no external AC input. Instead, a *resonant circuit* consisting of a fixed or variable capacitor in parallel with a *tapped inductor* is connected across what would be the input in an amplifier. (Compare Figures 4.1 and 4.2 with Figures 2.1 and 2.2 in chapter 2.)

Part of the inductor is connected between the cathode or emitter and "ground" (B–), or between the plate and collector and ground, to provide *positive inductive feedback*, as in Figures 4.1 and 4.2.

Uses

Hartley oscillators are used in a great many different circuits, including radio receivers and transmitters, audio oscillators and other applications where a sine wave signal is required.

Detailed Analysis

DC Subcircuits: In Figure 4.1 electron flow is from B–, via the lower portion of L1 to the tap, to the cathode of V; and from the plate of V back to B+.

In Figure 4.2 electron flow is from B– to the emitter of Q; and from the collector via L2 back to B+.

AC Input Circuits: In Figure 4.1, when power is applied current flows through the lower portion of L1 to the cathode of V, increasing as it overcomes the coil's inductive reactance. The changing flux lines

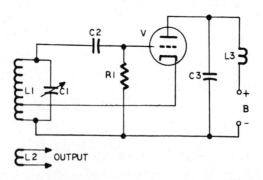

Figure 4.1 Hartley Oscillator (Cathode Feedback)

Oscillators

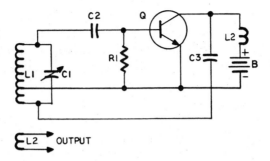

Figure 4.2 Hartley Oscillator (Collector Feedback)

induce a similar build-up of current in the upper portion of L1, which charges positively the upper side of C1. This increasing positive voltage is coupled to the grid of V by C2.

A positive bias on the grid of V causes the tube to conduct heavily, resulting in a further increase of current in L1, and even higher voltage on the grid of V.

The positive voltage on the grid would continue to rise indefinitely except that it increasingly attracts electrons in transit from the cathode to the plate, until they build up a negative charge that balances the positive voltage. The current through the tube levels off, and consequently the current through the lower part of L1. No more is induced in the upper part of the coil, because the flux has stopped changing. C1 now starts to discharge. Its positive charge dissipates faster than the negative charge on the grid, so the tube is gradually cut off, and ceases conducting.

C1 is discharging through L1. The excess of electrons on its lower side flows up through L1 to neutralize the positive charge on the upper side. However, as the nature of a coil is to oppose any change in the current flowing through it, this current goes on flowing even after C1 is completely discharged, until the upper side of C1 becomes negatively charged, reinforcing the negative bias on the grid of V. Eventually the negative charge on the upper side of C1 builds up sufficiently to stop the current flowing through L1, and now the excess of electrons begins to flow in the opposite direction. This continues until the negative charge has been returned to the lower side of C1 and a deficiency of electrons makes the upper side positive again.

The restoration of a positive charge to the upper side of C1 is coupled by C2 to the grid of V, and starts the tube conducting again, so that the complete cycle is repeated. This goes on, cycle succeeding cycle, as long as power is applied to the circuit.

The frequency of oscillation depends upon the reactance of L1 and the capacitance of C1, as their combination determines the rate at which the charge of electrons is switched back and forth between the two sides of C1.

In Figure 4.2 the only difference is that the lower portion of L1 is in series with the collector, so that the feedback is from the output side of the transistor, and is therefore of opposite phase to the input. The tap on L1 is therefore connected to the low side of the circuit instead, so as to have the feedback in proper phase to reinforce the oscillations in the grid.

AC Output Subcircuits: The current flowing to and fro in L1 builds up and dies down in such a way that the voltage across it alternates between positive and negative in the smooth curve we call a sine wave. These sine waves are coupled inductively into the output coil L2.

COLPITTS OSCILLATOR

The Colpitts oscillator is very similar to the Hartley oscillator, except that it uses capacitive feedback instead of inductive feedback. Figures 4.3 and 4.4 are examples of this circuit.

Distinguishing Features

The Colpitts oscillator employs a vacuum tube or transistor connected in a *grounded-cathode* or *common-emitter* circuit.

There is no external input. Instead, a resonant circuit consisting of an inductor in parallel with *two capacitors in series* (or a split-stator variable capacitor, or the equivalent), which is the capacitive opposite number of the Hartley tapped inductor, is connected in a similar way to give *positive capacitive feedback*.

Uses

Colpitts oscillators are used in a great many different circuits, including radio receivers and transmitters, audio oscillators and other applications where a sine wave signal is required.

Oscillators

Detailed Analysis

DC Subcircuits: In Figure 4.3 electron flow is from B– via V and L2 to B+. In Figure 4.4 it is from B– via Q, R3 and L2 to B+. Bias for Q is stabilized by R1 and R2.

AC Subcircuits: In Figure 4.3, when the oscillator is first turned on current flows through L2, causing a voltage drop which charges C3. The charge on C3 starts oscillation in the resonant circuit consisting of L1, C1, C2 and C3. (As far as resonance is concerned the three capacitors count as one.) The voltage developed across C2 is the feedback voltage, which is applied between the grid and cathode of V. As this voltage increases positively it causes the resistance of V to decrease, so that it conducts more. This, in turn, increases the current

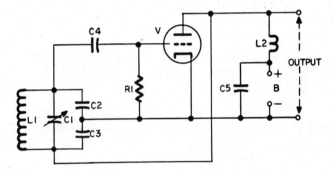

Figure 4.3 Basic Colpitts Oscillator

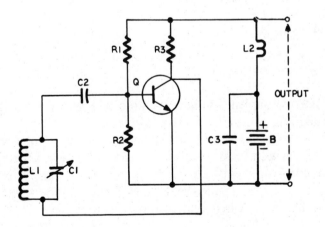

Figure 4.4 VHF Colpitts Oscillator

flow through L2, and charges C3 more. Eventually the electrons collecting on the grid of V balance the increase in the positive voltage, as you saw in the Hartley oscillator, and C2 starts to discharge through L1.

The amount of feedback depends upon the ratio of the values of C2 and C3. The smaller capacitor will have the higher reactance, and will have the greater voltage drop across it. If this is C2, a higher voltage will be applied between grid and cathode; if C3, a lower voltage.

C1 is a variable capacitor, which allows the resonant circuit to be tuned to the desired frequency. If it is omitted, the resonant circuit will oscillate at a single frequency determined by the combination of L1 and C2 and C3 (which will act as a single capacitor).

In Figure 4.4, the capacitors C2 and C3 of Figure 4.3 have been omitted, but otherwise the circuit is the same. At the higher frequencies used in TV and FM tuners the base-emitter capacitance of Q replaces C2, and the collector-emitter capacitance of Q replaces C3. If you were to sketch in these capacitances you'd have the circuit of Figure 4.3 again, except for a transistor instead of a vacuum tube.

The circuit in Figure 4.4 is also known as the Ultra-Audion oscillator. When it was invented it was not immediately recognized for what it was — a VHF Colpitts oscillator. It doesn't matter which name you use.

ELECTRON-COUPLED OSCILLATOR

Figure 4.5 shows an electron-coupled oscillator.

Distinguishing Features

The electron-coupled oscillator employs a *pentode or tetrode* vacuum tube in a *grounded-cathode* circuit.

There is no external input. Instead, a Hartley or Colpitts resonant circuit is connected across what would be the input in an amplifier.

The vacuum-tube cathode, control grid and screen grid are connected as if they were the cathode, grid and plate of the triode in Figure 4.1. The output is taken from the pentode plate.

Uses

Triodes provide considerable capacitive coupling between input and

Oscillators

output. In some cases this can be a problem, as, for example, when loading the output causes a frequency shift in the resonant circuit in the input. In circuits subject to this, and where it would be a disadvantage, the isolation between input and output provided by a pentode or tetrode vacuum tube in an electron-coupled oscillator is one answer.

Detailed Analysis

DC Subcircuit: In Figure 4.5 electron flow is from B− via the lower portion of L1 to the cathode of V, and thence from the plate via L2 and the screen grid via R2 back to B+.

AC Subcircuits: The circuit in Figure 4.5 is a Hartley oscillator, identical with that in Figure 4.1, except that the output is taken from the plate of V. The cathode, control grid and screen grid of V act as a triode which is kept cut off most of the time by the negative voltage on the control grid. When the grid goes positive a strong current pulse flows through the tube, causing output signals from both screen grid and plate. The screen-grid signal provides positive feedback via C3 to maintain oscillation, the action of this part of the circuit being exactly as described previously for the Hartley oscillator. In the plate circuit the output signal voltage across L2 is coupled to the next stage. A resonant circuit similar to that in Figure 4.6 may replace L2.

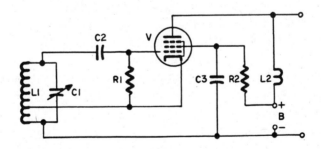

Figure 4.5 Electron-Coupled Oscillator

TUNED-PLATE, TUNED-GRID OSCILLATOR

Figure 4.6 illustrates a tuned-plate, tuned-grid oscillator, so called because there are tuned resonant circuits in both input and output circuits.

Distinguishing Features

A triode or transistor is connected in a *grounded-cathode* or *common-emitter* circuit.

There is no external input. Instead, a tuned resonant circuit is connected across what would be the input in an amplifier.

A *tuned resonant circuit in the plate or collector circuit* provides the means of coupling the signal into the next stage.

Uses

This circuit is not used as much as the Hartley or Colpitts circuits, but you will meet it frequently in the form of a *crystal oscillator* (described next).

Detailed Analysis

DC Subcircuit: In Figure 4.6, electron flow is from B– via V, L2 and L3 back to B+.

AC Subcircuits: When power is turned on electrons flow through the tube and into the side of C3 connected to the plate. At the same time, current begins to flow through L2. The resonant circuit C3-L2 begins oscillating, and at the correct moment a positive voltage pulse is fed back to the grid, through the plate-grid capacitance.

The effect of this feedback voltage is to drive the grid in a positive direction, which increases current through the tube and charges C3 still further. However, the positive grid now attracts electrons from the stream of electrons passing from cathode to plate, and they start to flow back to the cathode via R1. This resistor has a high resistance, so

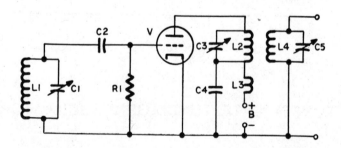

Figure 4.6 Tuned-Plate, Tuned-Grid Oscillator

Oscillators

at this stage electrons build up on the side of C2 connected to the grid faster than they can leak away through R1.

When the positive feedback signal was swinging the grid in a positive direction it was also coupled through C2 to C1, and caused a positive charge to appear on the upper side of this capacitor. But as electrons build up on C2 the grid gradually becomes negative and the tube ceases to conduct. When this happens the feedback voltage disappears and C1 starts to discharge through L1. This causes the resonant circuit C1-L1 to start oscillating at a frequency determined by the inductance of L1 and the capacitance of C1.

Meanwhile C2 completes discharging through R1, and the grid loses its negative voltage, so that V can conduct again. The resonant circuit C3-L2 gets another current pulse to sustain oscillation, and in turn sends one back through V as before, so that the same cycle is repeated continuously as long as power is applied to the circuit.

C3 has to be adjusted so that the frequency of C3-L2 is slightly lower than that of L1-C1 if the feedback voltage is to be in proper phase. The oscillations of the output circuit are coupled into the next stage via L4 and C5, which also form a tuned resonant circuit.

CRYSTAL OSCILLATORS

Figure 4.7 illustrates a crystal oscillator, which is identical to a TPTG oscillator, except that a crystal replaces L1-C1-C2 in Figure 4.6.

Distinguishing Features

A triode or transistor is connected in a *grounded-cathode* or *common-emitter* circuit.

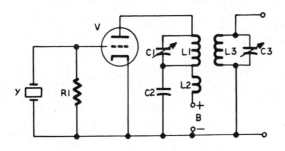

Figure 4.7   Crystal Oscillator

There is no external input. Instead, a *crystal* (Y) is connected across what would be the input in an amplifier.

A tuned resonant circuit in the plate or collector circuit provides the means of coupling the signal into the next stage.

There is no visible feedback path.

Uses

Crystal oscillators are much more stable than other oscillators; consequently they are preferred where frequency must be controlled to close tolerances. Examples are found in radio transmitters operating on fixed frequencies (such as broadcast stations), precision signal generators, counters and similar applications.

Detailed Analysis

The operation of the circuit in Figure 4.7 is identical with that in Figure 4.6. The crystal (Y) replaces the resonant circuit in the grid circuit of the TPTG oscillator.

A quartz crystal will vibrate mechanically at a frequency depending upon its thickness, therefore this frequency is fixed by physical considerations only. In the oscillator the feedback pulses excite mechanical vibrations in the crystal which in turn generate electrical vibrations exactly as if the crystal were a resonant circuit consisting of inductance, capacitance and some resistance. This interaction between mechanical and electrical functions in a crystal is called the *piezoelectric effect.*

As the dimensions of the crystal are affected by temperature, the frequency is temperature-sensitive. For very accurate frequency control, crystals are installed in *crystal ovens,* in which they are maintained at a constant temperature by a thermostat.

Circuit Variations

A crystal can replace an L-C circuit in other types of oscillator. For example, Figure 4.8 shows how this can be done with the Colpitts oscillator. Compare this circuit with the one shown in Figure 4.3. This circuit could also have a FET (field-effect transistor) instead of a bipolar transistor.

Oscillators

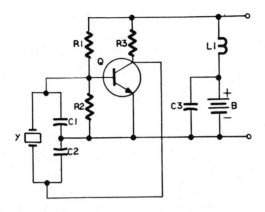

Figure 4.8 Crystal Oscillator (Colpitts Type)

ARMSTRONG, OR TUNED-GRID OSCILLATOR

Figure 4.9 shows an Armstrong oscillator, with the tuned circuit in the grid circuit.

Distinguishing Features

A vacuum tube or transistor is connected in a *grounded-cathode* or *common-emitter* circuit.

There is no external input. Instead, a tuned resonant circuit is connected across what would be the input in an amplifier.

A feedback circuit inductively couples the output signal to the grid resonant circuit.

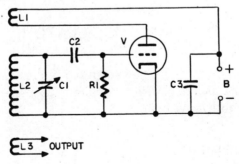

Figure 4.9 Armstrong or Tuned-Grid Oscillator

Uses

This circuit is not used as much as the Hartley or Colpitts circuits, but is very similar to the popular blocking oscillator circuit discussed next, to which it makes a good introduction.

Detailed Analysis

DC Subcircuit: In Figure 4.9 electron flow is from B− via V and L1, back to B+.

AC Subcircuit: The operation of this oscillator is similar to that of the TPTG oscillator, except that feedback is by means of inductive coupling between L1 and L2 instead of via the interelectrode capacitance of the tube. When power is turned on, the plate current starts to flow through L1, building up a magnetic field which induces a voltage in L2. This induced voltage charges C1, and the resonant circuit begins to oscillate in the same way as in the Hartley oscillator discussed at the beginning of the chapter.

R-C (RELAXATION) OSCILLATORS

Another name for an R-C oscillator is relaxation oscillator. Any oscillator in which the frequency is determined by the rate at which a capacitor can be charged and discharged through a resistance is an R-C or relaxation oscillator. Multiplying the value of the resistance (in ohms) by the value of the capacitance (in farads) gives the time constant or period, and its reciprocal is the oscillator frequency.

BLOCKING OSCILLATOR

A blocking oscillator is a well-known type of relaxation oscillator. Figure 4.10 illustrates a blocking oscillator as used in a TV set to generate a sawtooth waveform.

Distinguishing Features

A triode or transistor is connected in a *grounded-cathode* or *common-emitter* circuit.

Oscillators

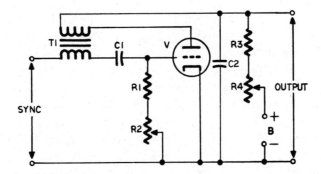

Figure 4.10 Blocking Oscillator (Vacuum Tube)

The AC input subcircuit contains a *transformer* by which feedback from plate circuit to grid circuit is obtained. You might possibly think that this makes it an L-C oscillator, but another look at the transformer will show that there is no parallel capacitor to tune it to resonance, so it is not part of a resonant circuit.

There is usually an external input (a *synchronizing signal*) which is another indication that this is different from oscillators discussed previously.

Uses

Blocking oscillators are used in many TV circuits to provide the driving signals for vertical (field) and horizontal (line) deflection in the picture tube. These signals take the form of a sawtooth, in which a steady linear rise in voltage is followed by a sharp drop-off.

Detailed Analysis

DC Subcircuit: In Figure 4.10 electron flow is from B− via V, the secondary winding of T1, R3 and R4, to B+.

AC Subcircuits: When power is first applied to this circuit an increasing plate current flows through the upper winding of T1, and because the magnetic field is building up it induces a voltage in the lower winding, with a polarity that couples through C1 to make the tube grid positive. This attracts electrons to the grid, and a negative charge begins to build up on C1 until V is cut off. It doesn't take long to charge C1 from the grid, because the tube internal resistance is low between the cathode and grid when the latter is positive. However, when the tube cuts off it is as if a switch had opened. The only way for C1 to discharge is through R1 and R2. But these resistors have comparatively high values, so it takes C1 considerably longer to discharge than it did to charge. This means that the time during which the tube is conducting is much shorter than the time when it is not. In short, the tube is an automatic switch which is normally "off," but which turns "on" briefly to allow short bursts of current.

So far the circuit has been performing somewhat like the Armstrong oscillator in the previous section. But something different is happening in the plate circuit. While the tube is "turned off" the B voltage starts to charge C2 by pulling electrons away from its upper side. The values of C2, R3 and R4 are chosen so that C2 will not charge to more than about a tenth of the B+ voltage before the tube conducts again. When this happens the burst of electrons supplied by the current pulse through V wipes out the positive charge on C2 (by restoring the missing electrons), and the voltage drops back to its initial value. The result is a sawtooth output voltage, with a slower steady rise and a faster, sudden fall. Its amplitude is adjusted by R4, which in a TV set would be the height control in a vertical deflection circuit, or the width control in a horizontal deflection circuit.

The frequency of this sawtooth output signal is determined by the time constant of C1, R1 and R2. Since R2 is variable it can be used to vary the frequency of the oscillator. In a TV set this variable resistor is the hold control.

Without a synchronizing signal the oscillator would free-run at its own frequency, which would not be exactly that of the TV signal, so you would get a picture that drifted or rolled. A sync signal is therefore introduced as shown. In a vertical deflection circuit the sync signal will consist of a sequence of positive pulses at the vertical sync rate (60 hertz). On arrival at the grid each positive pulse will cause the tube to turn on, and the circuit will perform one cycle of oscillation and one sawtooth, provided that R2 has been adjusted so that the grid cycle is almost at the point where the tube would have turned on of its own accord.

In a horizontal deflection circuit the sync signal frequency is compared with the oscillator frequency in a *frequency comparator* circuit (see chapter 12). The frequency difference becomes a DC voltage. If the oscillator is too slow the voltage will be positive, if too fast it will be negative. This voltage is applied to the grid of the tube or base of the transistor. A positive voltage will decrease the negative voltage on C1, so that it will discharge faster, speeding up the

Oscillators 131

frequency of oscillation, while a negative voltage will have the opposite effect. As the DC voltage is proportionate to the frequency difference, the frequency will be adjusted automatically to that of the sync signal.

Circuit Variations

Transistor blocking oscillators work on the same principle as vacuum-tube blocking oscillators, with some variations due to the different characteristics of transistors. In Figure 4.11 you see an example of one used in solid-state television sets. The feedback is from collector to base, as in the vacuum-tube circuit you've just seen. The free-running frequency is determined by C2, R1 and R2 in the same way as C1, R1 and R2 in Figure 4.10. Diode D1 is provided to protect the transistor from damage when it cuts off. When this happens the sudden collapse of the magnetic field in T1 generates a sharp voltage pulse of opposite polarity which is high enough to cause the transistor to break down if not suppressed. The diode is connected with its polarity such that it has no effect on the positive feedback or sync pulses, but short-circuits the reverse spike which we don't want. R4 and C3 are the sawtooth-forming combination, operating in the same way as R4, R3 and C2 in Figure 4.10. The height control in this circuit is located elsewhere. (You will often find it in the following stage instead of where we show it in Figure 4.10.)

In some transistor circuits the feedback is from emitter to base

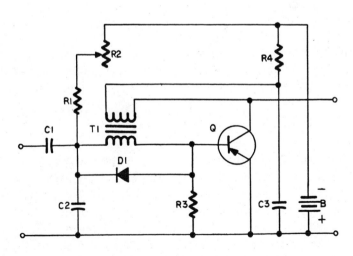

Figure 4.11 Blocking Oscillator (Transistor)

instead of from collector to base. Also, a PNP transistor would perform just as well, provided that we reversed the voltage polarity, including that of the sync pulses.

ASTABLE OR FREE-RUNNING MULTIVIBRATOR
(Monostable or Bistable Multivibrators are covered in chapter 12)

Astable multivibrators are also R-C or relaxation oscillators. There are two principle types, the *plate-coupled* (Figure 4.12) and the *cathode-coupled* (Figure 4.13).

Distinguishing Features

Two vacuum tubes (or sections of a dual tube) or *transistors,* each connected in a *grounded-cathode* or *common-emitter* circuit, are interconnected in either of the following ways:

1. *Plate-Coupled:* The plate of each tube is connected via a capacitor to the grid of the other (in the case of transistors, read collector and base, of course).
2. *Cathode-Coupled:* The cathodes of each tube are connected to each other, and share a common cathode resistor. The plate of V1A is connected via a capacitor to the grid of V1B, but not vice-versa.

External input of a synchronizing or triggering pulse or DC control voltage is usual, as in the case of blocking oscillators.

Since a multivibrator circuit looks very like a flip-flop, check under flip-flop in chapter 12 if you have any doubt about which it is.

Uses

Multivibrators are used in TV sets in the same way as blocking oscillators. They are also used in many industrial, military and space applications where a pulse output is required. Such uses are legion, but examples would include: timing pulse trains for multiplexing or time-sharing requirements for multi-channel communication and computer-controlled processes; cybernetic controls in guidance systems; sampling systems; and many more.

Detailed Analysis

DC Subcircuits: In Figure 4.12 electron flow is from B– through V1A and R1, and V1B and R4, back to B+. In Figure 4.13 electron flow is from B– to R3, and thence through V1A and R1, and V1B and R5, back to B+.

AC Subcircuits of Plate-Coupled Multivibrator: The circuit of Figure 4.12 is symmetrical. Each component in one half is duplicated in the other. It is a double switch, in which the tubes turn each other on and off alternately.

Actually, if the circuit were perfectly symmetrical it would be impossible to get it started, but as nothing is ever that perfect one tube will always begin to conduct before the other. For example, suppose V1A is just a fraction faster than V1B. As current begins to flow through R1 the voltage on V1A's plate starts to fall, and a negative-going pulse is coupled through C1 to the grid of V1B. This negative bias keeps on increasing as the current through V1A and R1 builds up, so that by the time V1A is conducting fully V1B is completely cut off.

However, when V1A's current levels off the negative charge on V1B's grid begins to leak away through R3 until V1B can conduct. When this happens its plate voltage falls, because of the potential drop across R4, and the negative-going voltage coupled to V1A's grid now starts to turn it off. This process continues back and forth as long as power is applied to the circuit.

If we just consider what is happening on C2, we can see that when V1B is turned off no current flows through R4, so the side of C2 connected to V1B's plate charges up to the full B+ voltage. When V1B turns on, the voltage drops by the amount dropped across R4. Since V1B is switching on and off continuously we get an output from C2 which alternates between these two voltage levels.

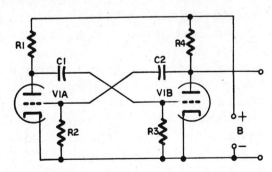

Figure 4.12 Plate-Coupled Multivibrator

The resulting signal is a symmetrical *square wave*, in a symmetrical multivibrator, but if the R-C combination C1-R3 has a different time constant to C2-R2 the duration of V1A's on and off times will be different from V1B's. This will give an asymmetrical output, in which the upper and lower voltage levels will not have the same duration.

AC Subcircuits of Cathode-Coupled Multivibrator: In Figure 4.13 the two halves of the multivibrator are not symmetrical as far as their components go, since there is only one coupling capacitor, C1. The action of this circuit is more like that of the blocking oscillator, where V1B is the tube and V1A the transformer. Let's see how this is.

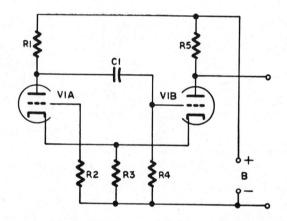

Figure 4.13   Cathode-Coupled Multivibrator

Current flowing through V1B causes a voltage drop across R3, which is also V1A's cathode resistor. This voltage drop across R3 biases V1A's grid negatively, so that conduction in V1A decreases and its plate voltage rises. This rising voltage is coupled through C1 to V1B's grid, increasing conduction in V1B. This, in turn, increases the voltage drop across R3, and the bias on V1A's grid, so that eventually V1A is cut off altogether.

However, the increasing positive voltage coupled through to the grid of V1B disappears as soon as V1A cuts off, because C1 can only pass a *changing* voltage, not a DC voltage, and is replaced by a negative bias resulting from conduction through R3. This leads to a downward swing in the conduction rate through V1B, which reduces the voltage drop across R3, and the bias on V1A's grid. V1A begins to conduct and its plate voltage starts to fall. This negative-going change in its plate voltage is coupled through C1 to V1B's grid, until the tube is turned off.

Oscillators

V1B remains turned off until the negative charge on its grid leaks away through R4. When it gets down to the level where V1B starts to conduct again a new cycle begins.

The output of this circuit is the same as in the plate-coupled multivibrator, a two-level DC voltage, the levels changing as V1B switches on and off. This DC signal becomes an AC signal by coupling it through a capacitor to the next stage.

SAWTOOTH GENERATOR

By connecting a capacitor and resistor in series across the output, as was done in Figures 4.10 and 4.11, the output signal would become a sawtooth. The operation of the circuit would be the same as was explained for those circuits.

Sync Input

Sync input is similar to that for blocking oscillators, and works in the same way.

Circuit Variations

Transistors frequently replace tubes in substantially similar circuits. There are also multivibrators in which a triode and a pentode replace the two triodes. Such a circuit would be one in which the second stage had, for reasons of economy, to double as a power output amplifier.

Such a circuit is shown in Figure 4.14. The tubes shown are often half-sections of a dual tube.

At first you might not see much similarity between Figure 4.12 and Figure 4.14, but this is only because we have shown the entire circuit, in which V2 is also a power amplifier driving the vertical deflection coils of the picture tube. The plate of V1 is connected to the grid of V2 through C6, and the plate of V2 is connected back to the grid of V1 through R8, C5, R3 and C2. It is only this feedback loop that complicates the picture.

The output signal from the plate of V2 is far too strong to apply directly to the grid of V1, and must also be filtered to get rid of nearly all the sawtooth, not to mention some strong spikes on it, before it can be used to control V1. The resistors and capacitors through which it passes, together with R4 and C3, form a filter. You can read about

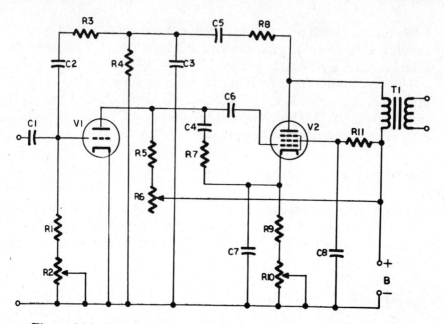

Figure 4.14 Plate-Coupled Multivibrator for TV Vertical Deflection

filters in chapter 8, so we'll not take time to go into it here. There are several variations possible in the arrangement of the resistors and capacitors in the filter, but they all serve the same purpose, to transform a portion of V2's output into a form suitable for application to V1's grid.

The sawtooth is formed by the charging and discharging of C4, and its amplitude is controlled by R6 (the height control), so this part of the circuit is the same as C2, R3 and R4 in Figure 4.10. Similarly, R2 is the hold control. The other control, R10, is the linearity control. It varies the DC bias on the grid of V2 in the same way as the control shown in the video amplifier in Figure 2.12. It allows you to shift the operating point of the tube to the proper point on its characteristic curve to give a linear sawtooth output. (See Appendix for explanation of vacuum-tube characteristic curves and class of operation.)

T1 is the vertical deflection output transformer, R11 the screen-grid voltage-dropping resistor, and C8 the screen-grid bypass capacitor.

Figure 4.15 shows a transistor collector-coupled multivibrator that operates in the same way as that in Figure 4.12.

Another type of sawtooth oscillator is described in chapter 12.

FETs (field-effect transistors) may be used in oscillator circuits also instead of tubes or bipolar transistors.

Oscillators

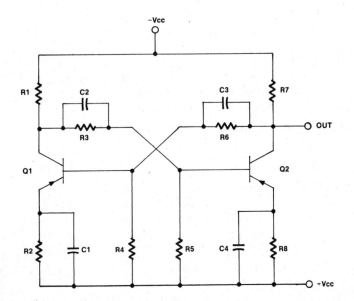

Figure 4.15 Collector-Coupled Multivibrator with Transistors. The base bias voltage for Q1 is stabilized by R4, R6 and R7; and for Q2 by R1, R3 and R5.

SIMPLE RELAXATION OSCILLATOR

Figure 4.16 shows the simplest form of relaxation oscillator.

Distinguishing Features

Capacitor and resistor in series with power source, shunted by a gas-filled tube or lamp.

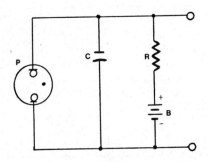

Figure 4.16 Simple Relaxation Oscillator

Uses

A simple relaxation oscillator is not stable or linear enough for TV circuits, but has practical applications elsewhere. In simple oscilloscopes it has been used to generate the sawtooth sweep signal, and it can be used also to make a light flash at a chosen frequency, as in a strobe lamp or the warning lamps used in road repairs.

Detailed Analysis

In Figure 4.16 the battery, B, charges C through R at a rate depending upon the time constant C x R. The gas in P ionizes when the voltage across C (which is also across P) reaches a certain critical level. P then becomes a low resistance, discharging C rapidly. The lamp then reverts to its original condition of high resistance, and C begins to charge again.

Circuit Variations

By using a *discharge tube* instead of a neon lamp a sync signal can be applied to its grid to trigger the tube, thus synchronizing the rate of discharge of C to the external source.

A *unijunction transistor oscillator* is a similar circuit, where the UJT is a solid-state equivalent of a discharge tube, as shown in Figure 4.17, where C1 and R2 are the sawtooth-forming combination. When the voltage on C1 reaches a level high enough to trigger Q, the resistance between the emitter (arrow) and base 1 (connected to R1) drops, and C1 discharges through Q and R1. The voltage on the emitter now is

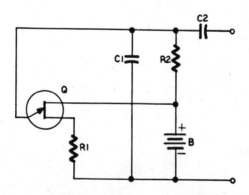

Figure 4.17   UJT Relaxation Oscillator

Oscillators

back where it was, the UJT turns off, and C1 starts to charge again.

A sync signal applied between the emitter and the low side of the circuit will control the rate of firing, as with a discharge tube.

This oscillator can also give a pulse output similar to that of a multivibrator (a rapid switching between two voltage levels) by taking the output from across R1 instead of C1.

TROUBLESHOOTING TABLE FOR OSCILLATORS	SYMPTOM						
	Oscillation absent or weak	Oscillation Intermittent	Oscillates at Wrong Frequency	Distorted output			
Vacuum Tube	X	X		X			
Transistor	X	X					
Coupling Capacitor	X	X					
Bypass or Decoupling Capacitor				X			
Defective Resistor	X	X					
Broken Wire or Defective Insulation	X	X					
Defective Control	X	X					
Misadjusted Control	X		X				
Dirty Contacts Anywhere	X	X					
Incorrect Operating Voltages	X			X			
Defective Power Supply	X			X			
Feedback Phase Reversed	X						

5

MODULATION

Since video and audio signals cannot be propagated through space they must be carried by radio signals that can. Adding audio or video information to a radio "carrier wave" is called *modulation*. In this chapter we shall discuss circuits used in *amplitude modulation (AM)*, *frequency modulation (FM)* and *phase modulation (PM)*, all of which are used in color television transmission. *Pulse modulation*, which is used in *telemetry*, will also be explained.

The modern radio and TV receiver is a *superheterodyne receiver*, in which the incoming modulated radio carrier is mixed with another radio-frequency signal generated in the receiver itself, and converted to an *intermediate frequency (IF)*, which is then amplified in the *IF amplifier* (chapter 2). Since this is also modulation, we include mixers and converters in this chapter.

AMPLITUDE MODULATION

In AM we wish to vary the power of the transmitted signal in accordance with audio information, so we need a circuit in which the output of an RF amplifier is controlled by an audio signal. In such an amplifier the carrier signal generated in a previous stage is amplified by the addition of power by means of a vacuum tube or transistor. The output-signal amplitude depends upon the type of circuit, tube

Modulation

characteristics and operating voltages. If one of the operating voltages is varied in accordance with an audio signal, the output power of the carrier will be varied proportionately.

PLATE MODULATION

In plate modulation, varying the plate voltage causes the DC current through the RF power amplifier tube to vary, thereby varying the output power. Figure 5.1 illustrates a typical circuit.

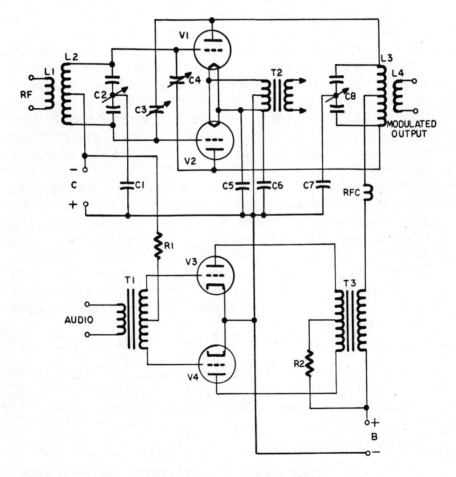

Figure 5.1   Plate Modulation

Distinguishing Features

The DC plate supply for the RF power amplifier comes through the secondary winding of a *modulation transformer* in the output circuit of an audio power amplifier. Both amplifiers may be single-ended or push-pull.

Uses

This circuit will be found in AM transmitters of all sizes.

Detailed Analysis

In Figure 5.1 the audio power amplifier is the same as those discussed in chapter 3, and its operation is identical. Similarly, the RF power amplifier is also the same as those described in the same chapter.

B+ current for the RF amplifier plate circuit has to pass through the secondary of T3. The audio signal from V3 and V4 is also present in the secondary of T3, and its variations in amplitude add to or subtract from the DC supply voltage. The resulting variations in the plate supply voltage cause corresponding variations in the gain of V1 and V2, so that the output amplitude of the carrier fluctuates in accordance with the audio signal.

The degree to which the carrier is modulated depends upon the power of the audio signal, and is expressed as a percentage. In 100 percent modulation the carrier is driven all the way to zero at its minimum amplitude, and to double its "resting" amplitude at the maximum value. This requires that the audio power amplifier be capable of as great an output as the RF power amplifier. This might seem a drawback, but plate modulation is much more efficient than grid modulation because the RF power amplifier can be operated Class C (see Appendix).

GRID MODULATION

Figure 5.2 shows a grid modulation circuit.

Distinguishing Features

The grid-bias supply voltage to the RF power amplifier is routed through the secondary winding of the modulation transformer in the

Modulation

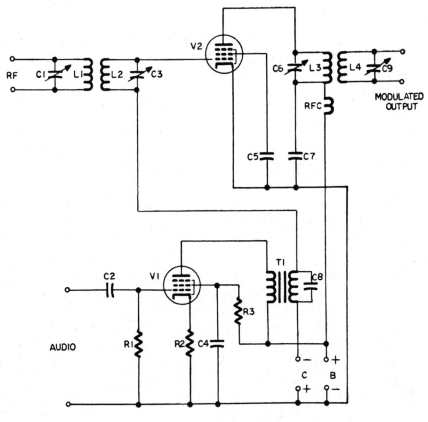

Figure 5.2   Grid Modulation

audio power amplifier output. Both amplifiers may be single-ended or push-pull.

Uses

This circuit will be found in AM transmitters of all sizes.

Detailed Analysis

In Figure 5.2 both amplifiers are similar to those discussed in chapter 3, so will not be analyzed in detail here.

C− (grid bias voltage) for the RF amplifier is routed through the secondary winding of T1. The audio signal from V1 is also present in the secondary of T1, and its variations in amplitude add to or subtract from the DC supply. The resulting variations in the grid bias voltage cause corresponding variations in the gain of V2, so that the output amplitude of the carrier fluctuates in accordance with the audio signal.

The secondary of T1 is shunted by C8. This capacitor's value is such that while it does not bypass T1 for audio it does for RF. In this way a connection is provided from the lower end of L2 via C8 and the grid-bias power supply to the cathode of V2.

In grid modulation the audio power amplifier does not have to be as powerful as the RF amplifier it modulates, since the output power must all come from the RF amplifier. This amplifier can only be operated Class B, however, since the audio information would be lost in Class C operation. In Class B operation (see Appendix) the positive swings of the modulating signal are completely reproduced in the tube. The cut-off negative swings are restored by the "flywheel" action of the output tank circuit C6-L3, as explained in chapter 4.

FREQUENCY MODULATION

In FM we wish to vary the *frequency* of the transmitted signal in accordance with audio information, so we need a variable frequency oscillator in which the frequency of the signal generated is not fixed, but is controlled by an audio signal. The variable-frequency carrier may then be amplified in subsequent RF amplifier stages as necessary for broadcasting.

Varying the frequency of an oscillator can be achieved in various ways. The two most well known are the *reactance circuit* and the *Armstrong balanced modulator*.

REACTANCE CIRCUIT

In Figure 5.3 a Hartley oscillator is shown controlled by a reactance circuit. The other sections of the FM transmitter are illustrated in block form.

Distinguishing Features

The Hartley oscillator (V2) is identical with that shown in Figure 4.1, so its distinguishing features need not be repeated here.

Modulation

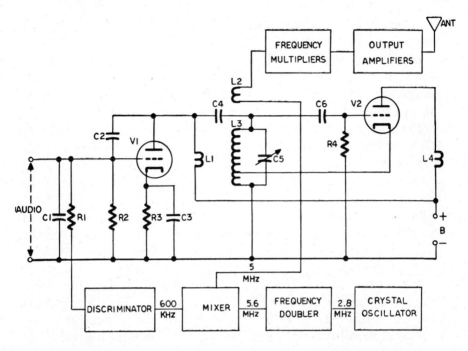

Figure 5.3 Reactance Circuit and Oscillator for FM Transmitter

The reactance circuit (V1) is similar to the audio voltage amplifier of Figure 2.1, except for the addition of C2 and the substitution of L1 for R2. An audio amplifier does not handle frequencies high enough to require an RF choke in its output. These two items, taken in conjunction with its position in the overall circuit, should make it difficult to mistake this circuit for an audio voltage amplifier despite its apparent similarity.

Uses

It is employed in FM transmitters.

Detailed Analysis

Analysis of the Hartley oscillator circuit is in chapter 4. Analysis of the reactance circuit is as follows:

DC Subcircuit: In Figure 5.3 electron flow is from B– via R3 to V1's cathode, and from its plate via L1 to B+. A DC voltage produced by the *discriminator* (see chapter 6) is applied to V1's grid via R1. R1 and C1 form a *low-pass filter* (see chapter 8).

AC Subcircuits: To L3 (the inductor in the oscillator resonant circuit) the entire reactance circuit looks like a capacitor. This is because the current in the circuit leads the voltage by 90 degrees, due to the predominant influence of C2. This capacitor has a capacitive reactance at the resonant frequency that is ten times the resistance of R2. Since the reactance circuit is equivalent to a capacitor connected in parallel to L3, any change in its apparent reactance will affect L3 as if a change in actual capacitance had taken place, and consequently the resonant-circuit frequency will change.

The apparent reactance of the reactance circuit includes the resistance of V1. But the tube's resistance varies with the signal on the grid. When an audio signal is applied, the tube's resistance varies with the audio voltage. Consequently, the apparent reactance of the circuit varies at an audio rate, with the result that the frequency of the oscillator resonant circuit varies accordingly. In this way the amplitude and frequency of the audio signal are impressed on the carrier as the amount of deviation and rate of deviation from the resting frequency of the FM carrier.

R3 is the cathode resistor, selected to give the required bias voltage for V1's grid as in an amplifier, and C3 bypasses it to prevent degeneration, as explained in chapter 2. L1 is an RF choke, to keep the carrier signal out of the power supply. C4 is a DC blocking capacitor to keep the power supply from being short-circuited through L3.

The modulated carrier is coupled inductively from L3 to L2, and goes via frequency multipliers (see chapter 12) and output amplifiers (see chapter 3) to the antenna. If the modulated carrier is 5 megahertz, as in Figure 5.3, the frequency multipliers will raise it to the radio frequency. In the FM broadcast band this is between 88 and 108 megahertz.

The carrier frequency of an FM broadcasting station must not be allowed to drift more than 2 kilohertz from its assigned frequency. This means that a broadcast frequency of 100 megahertz (approximately in the middle of the band) must be held to a tolerance of ±0.002 percent. Only a crystal-controlled oscillator can be this stable. Obviously, you cannot control a variable-frequency oscillator directly with a crystal as it would no longer be variable; so an indirect method must be used.

A separate crystal oscillator (see chapter 4) is provided to supply a very stable signal at a slightly higher frequency than the resting frequency of the Hartley oscillator. The two frequencies are combined

Modulation

in a *mixer* (discussed later in this chapter), and the *difference signal* is applied to a discriminator.

If the carrier begins to drift the discriminator converts the change in the difference-signal into a DC voltage, which changes the bias on the grid of V1 so as to bring the resonant frequency of the Hartley oscillator back to its proper value. Audio voltage feedback is filtered out of the DC voltage from the discriminator by C1 and R1.

Circuit Variations

A reactance can be made inductive instead of capacitive by substituting an inductor for C2. In this case the circuit will have inductive characteristics: the voltage will lead the current. The resonant frequency of the oscillator will vary with the varying inductance of the reactance circuit.

An inductive effect will also be obtained by interchanging R2 with C2, and making R2 ten times the value of C2 at the resonant frequency. The grid voltage now appears across C2, and because it lags the current, the plate current of V1 will lag the oscillator voltage, so that the reactance circuit will look like an inductor to L3.

Similarly, if an inductor is used in the place of C2, now between the grid and low side of the circuit, the reactance circuit will present a capacitive appearance to L3.

In the three variations just mentioned an additional DC blocking capacitor is required in series with the inductor or resistor connected between the plate and grid of V1. A large value capacitor is used so it will offer virtually no reactance to the resonant frequency, while keeping the plate voltage from reaching the grid.

ARMSTRONG BALANCED MODULATOR

Another method of FM is shown in Figure 5.4, which illustrates the Armstrong balanced modulator.

Distinguishing Features

The balanced modulator in the upper part of the schematic consists of a push-pull output stage which resembles an audio power amplifier (see chapter 3) with its output coupled to a single-tube audio voltage amplifier.

It differs from an actual audio push-pull amplifier, however, because the audio output transformer is replaced by the three coils L1, L2 and L3; and the screen grids of V1 and V2 are connected to opposite ends of the secondary of the modulation transformer T1. The audio input is connected to the primary. The output from an oscillator is connected to the input. (There may be an amplifier between the oscillator and the balanced modulator.)

Uses

It is used in FM transmitters.

Detailed Analysis

DC Subcircuits: In Figure 5.4 electron flow is as follows:

V1: From B– via R4 to the cathode; from the plate via R10 back to B+; and from the screen grid via the upper half of T1 and R9 back to B+.

V2: From B– via R5 to the cathode; from the plate via R11 back to B+; and from the screen grid via the lower half of T1 and R9 back to B+.

V3: From B– via R7 to the cathode; from the plate via R6 back to B+; and from the screen grid via R8 back to B+.

V4: From B– via R13 to the cathode; from the plate via R18 back to B+; and from the screen grid via R14 back to B+.

V5: From B– via R16 to the cathode; from the plate via R6 back to B+; and from the screen grid via R17 back to B+.

Grid bias for all tubes is determined by their cathode resistors.

AC Subcircuits: The master oscillator is a crystal oscillator with a tightly controlled frequency, typically of 200 kilohertz, so you can see these amplifiers will not be handling RF. The oscillator signal is applied to the input of V3 and also to the grids of V1 and V2. The input circuit of V3 is identical with that of similar audio voltage amplifiers described in chapter 2.

The oscillator signal is applied simultaneously to each grid of V1 and V2 without inversion; so as far as this signal is concerned the outputs of both tubes are in phase.

The audio input is obtained from an audio amplifier section in which the audio has been amplified and *pre-emphasized* (see next circuit in this chapter), hence this amplifier is called a *correction amplifier*. The

Modulation

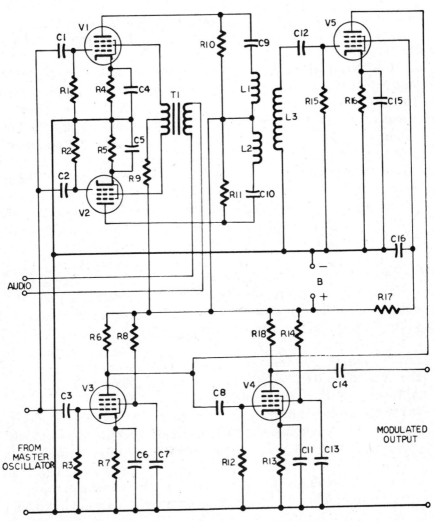

Figure 5.4 Armstrong Balanced Modulator

audio signal from the correction amplifier is applied to the screen grids of V1 and V2 via the transformer T1. The secondary of T1 is tapped, so that when the upper end of it swings positively in respect to the tap the lower end is swinging negatively; consequently the screen grids receive identical signals of opposite phase. Notice that these screen grids are not bypassed by capacitors, so the audio voltage appears on them and controls the plate current in each tube.

When no audio signal is present, equally amplified 200-kilohertz signals from V1 and V2 flow in opposite directions through L1 and L2, and because their induced magnetic fields are of opposite polarity they cancel each other, so no voltage is induced in L3 by the unmodulated carrier.

However, when an audio signal appears on the screen grids of V1 and V2 the tube receiving a positive-swinging signal will amplify more (because the grid will be more positive) than the other, with its negative-going signal, which will conduct less. Consequently, the 200-kilohertz signals appearing in L1 and L2 will be of unequal amplitude, and so a voltage will be induced in L3 by the stronger one. As this alternates between L1 and L2 the polarity of the voltage of the signal induced in L3 will alternate also.

The values of L1-C9 and L2-C10 have been chosen so that the currents in them are in phase with the voltage output from V1 and V2, but the voltage induced in L3 lags by 90 degrees.

This signal in L3 consists only of the sidebands, because, as you saw, the carrier by itself cannot induce any signal in L3. After amplification in V5's circuit they are recombined with the carrier at the output of V3. Because their phase is shifting back and forth at the audio rate, they cause the phase of the carrier to do the same.

However, when the phase shifts, the frequency shifts also, so the carrier signal that arrives at the grid of V4 is frequency modulated. V4's circuit is designed to *limit* the amplitude of the signal to get rid of any AM, some of which is bound to be present at this point.

The path followed by the signal from here to the antenna is similar to that in the reactance-modulator circuit described previously. However, in practical systems a *converter* stage will also be used, with a *local oscillator* which must be crystal-controlled as well. This is to overcome difficulties arising from using an excessive number of frequency-multiplication stages. Converter circuits are discussed later in this chapter.

PRE-EMPHASIS CIRCUIT

Before leaving this section on FM circuits we must consider the pre-emphasis circuit illustrated in Figure 5.5. In appearance this is an ordinary audio voltage amplifier. However, the value of C1 has been chosen to offer considerable reactance to lower frequencies. This reactance tapers off as the frequency increases, becoming virtually zero at the permitted upper audio limit of 15 kilohertz. As a result, the higher frequencies are emphasized, which gives a better signal-to-noise

Modulation

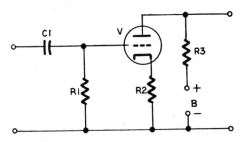

Figure 5.5   Pre-Emphasis Circuit

ratio in transmission. The value of R1 that gives the best performance is one that makes the C1-R1 time-constant 75 microseconds.

You will see that in FM receivers a corresponding de-emphasis circuit must be provided to restore the upper audio frequencies to their original level.

PHASE MODULATION

Phase modulation affects everyone who watches color television. The monochrome TV signal carries video by AM and audio by FM. The color signal includes color by PM.

In FM we vary the frequency of the carrier in accordance with the amplitude of the modulating signal, but in PM we vary its *phase angle*. This brings up the subject of *vectors*, which we must touch on briefly before going any further.

VECTORS

If we take a piece of paper and designate the top edge as north, and the left-hand side as west, and make a pencil dot in the middle for a ship, we can draw a vertical line from this dot to represent the ship traveling north, or a horizontal line to represent it traveling west. We have done this in Figure 5.6.

We can also represent the ship's speed. If we choose a scale of 10 knots to an inch, and the ship's speed is 20 knots, we shall draw the lines two inches long.

Suppose our ship is traveling northwest. Its course and speed are now represented by a two-inch line sloping up toward the left-hand upper corner of the paper, midway between the other two lines.

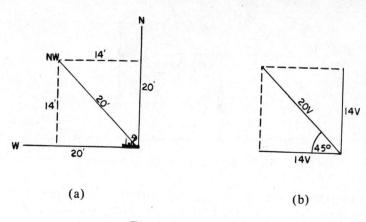

(a) (b)

Figure 5.6 Vectors

The point at the end of this line also shows that at the end of one hour the ship is 20 nautical miles northwest of its starting position. By measuring along the dotted lines, we can also say that the ship is now 14 miles north and 14 miles west of its starting point. Since it got there in one hour we could also say that the ship's speed was 14 knots to the north and 14 knots to the west, resulting in a total speed of 20 knots to the northwest.

In this diagram the three courses and speeds of the ship have been represented by lines drawn to show direction and velocity. Such lines are called *vectors*. In electronics, vectors are also used to show angles and voltages.

If, instead of compass headings and speeds, we substitute angles (measured clockwise from W) and voltages, we could use this diagram to show that two vectors, one of 14 volts at 0 degrees and one of 14 volts at 90 degrees, result in a vector of 20 volts at 45 degrees.

This is called *vector addition*. It is not the same as arithmetical addition, where 14 + 14 = 28, because the quantities added do not lie in the same straight line.

If they did they would be *in phase*. Where they do not the angle between them is called a *phase angle*.

COLOR MODULATION

In transmitting the color in color television a subcarrier is used. The space between the video and sound carriers in monochrome TV is

Modulation

occupied by the upper sideband of the video signal, which consists of groups of frequencies centered on harmonics of the line frequency, with spaces between. The color subcarrier has a frequency of 3.579545 megahertz (usually approximated to 3.58 MHz), which places it in the gap between the 227th and 228th harmonic of the line frequency. Harmonics of the color subcarrier fall into other spaces on either side of this one, since they are also separated by intervals equal to the line frequency (15.734264 kHz). In this way we can transmit a second signal interleaved within the main one.

This color signal has been modulated by shifting its phase according to the wavelength of the color. However, as we have just seen, a phase vector can be broken down into two vectors at right angles to each other. This is done by using a *doubly balanced modulator*.

DOUBLY BALANCED MODULATOR

Figure 5.7 shows a doubly balanced modulator that can be used to encode the 3.58-megahertz subcarrier with color information.

Distinguishing Features

You will notice a distinct similarity between this circuit and the Armstrong balanced modulator in Figure 5.4. However, the

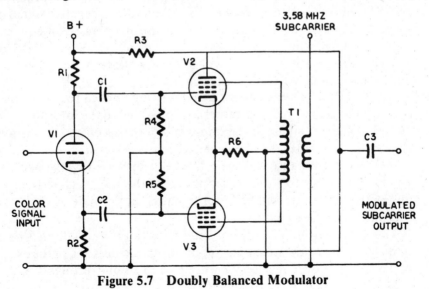

Figure 5.7   Doubly Balanced Modulator

modulating signal is actually fed to a phase-inverter triode (V1), from which *two* outputs of opposite phase go to the grids of V2 and V3. The carrier signal is applied to the third grids of these tubes, so the inputs are different from those of Figure 5.4.

Uses

TV color information encoding.

Detailed Analysis

DC Subcircuit: In Figure 5.7 electron flow is from B− (the low side of the circuit) via R2, V1 and R1 to B+; and via R6, both V2 and V3, and R3 to B+. Connections from the screen grids of V2 and V3 to B+ have been omitted.

AC Input : The color information derived from the red, green and blue cameras was changed by previous circuits into the Y, I and Q signals. The Y, being equivalent to the monochrome of the black-and-white pictures, bypasses this circuit. The I signal is fed to one of two doubly balanced modulators, the Q signal to the other.

The I signal is used to transmit colors in the orange and cyan regions, and since more fine detail is visible in these colors it is given a greater bandwidth (1.3 MHz). We will deal with this signal first.

When the I signal on the grid of V1 is positive, it will be reproduced by this tube as a positive signal on its cathode and a negative signal on its plate. (See chapter 3 for more about phase inverters.) Consequently two signals of opposite phase reach the grids of V2 and V3. V2's grid will go more negative and V3's will go more positive. V2 will therefore conduct less, and V3 will conduct more.

Meanwhile a 3.58-megahertz signal is being applied to the primary of T1. T1's secondary has a centertap connected to the low side of the circuit. When the signal at the upper end of the secondary is positive, that at the lower end will be negative, and vice-versa.

AC Output: Because V2's gain has been decreased by the negative voltage on its control grid, the amplitude of the 3.58-megahertz signal appearing at its plate will be lower than that at the plate of V3, where the gain has increased.

The output from V3 will not only be greater, but will also be of opposite phase to the subcarrier input. A signal consisting of the excess of V3's output over that of V2 will be coupled out through C3.

Modulation

The same action takes place when the input to V1 is negative, except that the output signal is now in phase with the subcarrier.

The Q Signal: In the other doubly balanced modulator the Q signal is applied to V1, and the operation of the circuit is identical. However, the phase of the subcarrier applied to this circuit is 90° different from the other. Consequently its output is also 90° different. The bandwidth of this signal is also less (.4 MHz), as colors other than those in the I signal have less visible detail.

In our brief discussion of vectors you saw how two vectors at a 90° angle would add to give a resultant vector. The angle of this vector would depend upon the values (in this case amplitudes) of the first two. This is illustrated in Figure 5.8, where the red, blue and green vectors are shown as the result of adding different I and Q modulation amplitudes. The outputs of the two doubly balanced modulators are actually combined in this way, and since their amplitudes are rising and falling, and varying between being in-phases and being of opposite phase to their subcarriers, you can see that the resultant vector is

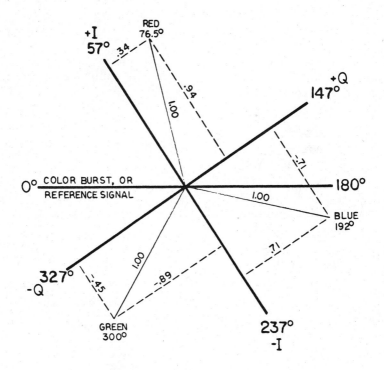

Figure 5.8 Color-Signal Phase Relationships

swinging back and forth around the center like the hand of a seriously deranged clock. Each direction it points to is a different color, of course, so that in this way the phase angle of the subcarrier is made to vary in accordance with the color information.

As already mentioned, more detail is visible in the orange and cyan portions of a scene than in those of other colors, and the I signal, with greater bandwidth, is used to transmit this part of the spectrum. Since the phase angle for orange is 57° and that for cyan approximately 180° the opposite, the I subcarrier is delayed by this angle before being applied to the I modulator. Consequently the Q subcarrier is delayed 147° (57 + 90) before being applied to the Q modulator. The color bursts, however, are not phase-shifted.

Don't forget that the Y signal bypasses the color modulators. Yet it contains the "whiteness" information for the picture. In the receiver it is recombined with the color signals so that the colors on the screen have the proper saturation as well as hue.

PULSE MODULATION

Instead of modulating a continuous sine-wave carrier, we can modulate a series of regularly recurrent pulses by causing the modulating signal to vary the amplitude, duration, time of occurrence, or grouping of these pulses, as shown in Figure 5.9.

In *pulse-amplitude modulation* (PAM), the series of periodically recurring pulses is modulated in amplitude by corresponding instantaneous *samples* of the modulating signal.

In *pulse-duration modulation* (PDM), the time of occurrence of either the leading edge or the trailing edge (or both) of each pulse is varied from its unmodulated position by the samples of the modulating signal. This is also called *pulse-length modulation* (PLM) or *pulse-width modulation* (PWM). A variation of PDM is also called *pulse-time modulation* (PTM).

In *pulse-position modulation* (PPM), the samples of the modulating signal are used to vary the position of a pulse relative to its unmodulated time of occurrence. This is also called *pulse-phase modulation* (PPM).

In *pulse-code modulation* (PCM), the modulating signal is sampled at regular intervals, and the amplitude digitized in the form of coded groups of pulses. A variation of PCM is called *delta modulation* (DM).

In *pulse-frequency modulation* (PFM), the repetition rate of the

Modulation

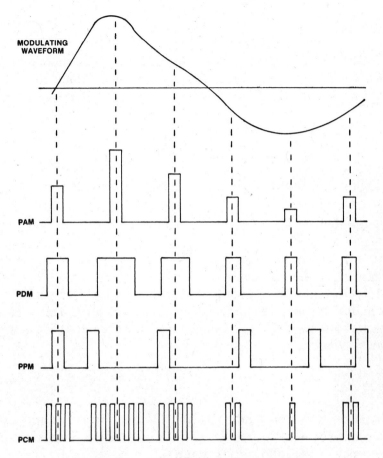

Figure 5.9 Types of Pulse Modulation

pulses varies in accordance with the amplitude of the modulating signal. It is, therefore, similar to FM.

Uses

Pulse modulation is used for *time-division multiplexing* (TDM). In TDM several pulse trains can be transmitted simultaneously via a single channel. They are interleaved in a similar manner to the black-and-white and color signals in a television channel (see above). A "clock" keeps them exactly synchronized so that they can be separated into individual pulse trains when they reach their destination. This is also called *time-sharing*. When the signals are transmitted over a considerable distance it is called *telemetering*.

Circuits

Pulses are generated by R-C oscillator circuits similar to those described in chapter 4. These "carriers" are then modulated as in AM, FM or PM, or "gated" by logic circuits (discussed in chapter 9).

MIXER CIRCUIT

Figure 5.10 shows a mixer circuit using a pentode tube.

Distinguishing Features

A mixer resembles other RF or IF voltage amplifiers (compare with Figure 2.22), having a *grounded-cathode* or *common-emitter* circuit.

It is distinguished from other RF or IF voltage amplifiers by having an additional input to the grid from an oscillator. This oscillator, which is often a Hartley or Colpitts type (see chapter 4), is known as a *local oscillator*.

Uses

A mixer is used in a radio receiver, TV tuner or similar application, in order to change the frequency of the incoming RF signal to the IF frequency of the following IF stage(s).

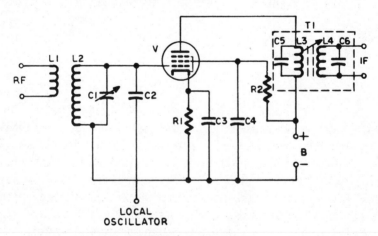

Figure 5.10 Mixer

Modulation

Detailed Analysis

DC Subcircuit: In Figure 5.10 electron flow is from B− via R1 to the cathode of V. From V's plate, current returns to B+ via L3, and from the screen grid via R2.

The value of R1 is chosen to bias the tube so that it operates on the non-linear portion of its characteristic curve, which makes it act like a detector. Consequently, in a radio receiver, this circuit is often called the "first detector." However, audio information is not removed here but at a later stage, sometimes referred to as the "second detector," which will be discussed in chapter 6.

AC Input Subcircuit: The resonant circuit L2-C1 is tuned to be resonant at the frequency of the radio or TV signal it is desired to receive. This signal is present in L1, is inductively coupled into L2, and gets a resonant-voltage step-up in the process. It is then applied between the control grid and cathode of V.

As you just saw, the tube is operating on the non-linear portion of its characteristic curve (see Appendix). Consequently, when the grid voltage swings in a negative direction the tube is cut off; it conducts only on the positive swings. This is similar to the operation of the RF amplifier used in the grid-modulation circuit discussed earlier in this chapter. However, that was a high-power amplifier, whereas this is a low-power voltage amplifier.

The reason for operating the tube near the cut-off point is because it is then able to modulate one signal with another. The oscillator signal, which is applied to the grid by way of C2, "beats" with the incoming RF signal to produce upper and lower sidebands that are equal to the sum and difference frequencies of the two signals. This process is called *heterodyning*.

If the incoming signal is at a frequency of, say 1070 kilohertz, and the local oscillator is operating at 1520 kilohertz, the upper sideband frequency will be 2590 kilohertz (1070 + 1520), and the lower sideband 450 kilohertz (1520 − 1070). If the incoming signal was already modulated by (for illustration) a single frequency of 1 kilohertz, it will have sidebands of its own at 1069 and 1071 kilohertz. These will give sidebands on sidebands: upper sidebands of 2589 and 2591 kilohertz; and lower sidebands of 449 and 451 kilohertz. Of course, in a real transmission there are many more sidebands, all of which will appear in the output with new upper and lower sidebands also.

We're only interested in the lower sidebands around 450 kilohertz. In AM broadcasting most stations are allotted a bandwidth extending

5 kilohertz above and below the carrier frequency. (A few stations are allowed up to 15 kilohertz, but there are not many of these "clear-channel" stations.) The IF transformer T1 is made resonant at 450 kilohertz, with a broad enough bandwidth to allow the 450-kilohertz carrier and its sidebands to pass, but excluding the rest. (See also section on IF amplifiers in chapter 2.) This type of radio receiver is called a *superheterodyne* receiver, and is almost the only type used today.

Circuit Variations

Mixers using transistors are found with circuits similar to that in Figure 5.10. They may also be found with dual tubes, containing a triode (for the oscillator) and a pentode (for the mixer) in one envelope. Such a tube might be employed in the next stage after the RF cascode amplifier shown in Figure 2.25.

CONVERTERS

Figures 5.11 and 5.12 are examples of converter circuits, in which mixer and local oscillator are combined in one circuit.

Distinguishing Features

  a. Vacuum-tube converters use *pentagrid tubes* (tubes with five grids, also called heptodes because there are seven electrodes).

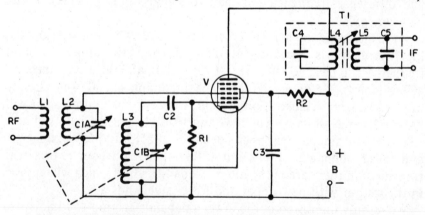

Figure 5.11   Pentagrid Converter

Modulation

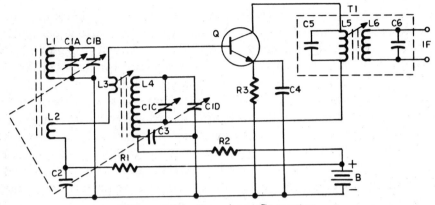

Figure 5.12 Transistor Converter

A pentagrid tube is enough to identify a converter. However, it is usually also connected in a circuit with a tunable resonant input which is ganged to track with the oscillator resonant circuit; and an IF transformer is provided in the output, as in the mixer circuit discussed previously.

b. Transistor converters are just as easy to identify, even though the transistors do not have five grids, since they are connected in the same kind of circuit. In the example given in Figure 5.12 a loop-stick antenna with a ferrite core (L1 and L2) is also shown, a feature of practically all transistor radios today.

Uses

Just about all AM broadcast receivers use converters today. They are also used in many other types of receiver, though sometimes preceded by an RF amplifier stage, as, for example, in TV receivers.

Detailed Analysis

DC Subcircuits: In Figure 5.11 electron flow is from B− via the lower part of L3 to the cathode of V. From the plate it returns through L4 to B+, and from grids numbers 2 and 4 via R2 to B+.

In Figure 5.12 electron flow is from the negative pole of the battery through R3 to Q's emitter, and from the collector via L5, the lower portion of L4 and R2, back to the positive pole of the battery. Bias is stabilized by R3 and R1.

AC Subcircuits: Each circuit has two inputs, the oscillator and the external RF. The oscillator in each case is a Hartley oscillator (sometimes a Colpitts is used). That in the tube circuit (Figure 5.11) is identical with the circuit in Figure 4.1. The transistor circuit differs a little, but you'll see why that is when we get to it.

In the pentagrid tube the grids are numbered from 1 through 5, starting from the one next to the cathode. Grids 2 and 4 are connected internally, so as to shield the third grid and minimize coupling between the circuits. Grids 2 and 4 are maintained at signal ground (AC zero) potential by C3. Grid No. 5 is a suppressor grid, and is connected internally to the cathode as shown, or externally to the low side of the circuit.

In Figure 4.1 the tube is a triode. In the pentagrid in Figure 5.11 the function of the triode is performed by the cathode, No. 1 grid and No 2 grid (No. 2 grid acting as the triode plate). The electron current flowing from the cathode to No. 2 grid is made to vary at the oscillator frequency by the oscillating voltage on the No. 1 grid. The operation of the Hartley oscillator is explained in chapter 4, so will not be repeated here. However, the No. 2 grid does not present a complete barrier to electrons. Attracted by the higher positive potential of the plate, most of them pass on through grids 2, 3, 4 and 5.

As they pass through grid No. 3 the stream of electrons is modulated by the varying RF signal voltage on this grid. This results in the appearance, in the output stage, of the oscillator and RF carrier signal frequencies, together with all the sidebands, as already explained in the section on the mixer circuit preceding this. T1 is an IF transformer tuned to let pass the IF frequency and its set of sidebands, while excluding the others.

The RF input may be from an external antenna or an internal loop antenna. In modern sets the ferrite loopstick L1 and L2 shown in Figure 5.12 is very popular for broadcast-band reception. It is tuned to resonance at the desired frequency by means of C1A, which is ganged with C1B (Figure 5.11). Usually a combined variable capacitor is used, in which both rotors share a common shaft, and turn together. This is to enable them to "track" or maintain a fixed difference frequency. The oscillator frequency must always be above that of the incoming RF frequency by the amount of the IF frequency, or there will be no difference frequency able to pass through to the IF stage.

C1A, C1B and L3 are always provided with trimmer capacitors and adjustable ferrite slugs for exact alignment for accurate tracking and tuning dial indication. These are shown in Figure 5.12 (C1A, C1C and L4), and are indicated in most schematics.

Modulation

In the transistor converter, apart from the features already mentioned, we see that the oscillator signal is coupled inductively to the base by L3 and L4 instead of capacitively by C2 as in Figure 5.11. You will meet both types of coupling in these circuits.

The main difference between the pentagrid converter and the transistor converter is that in the tube we can see the cathode and grids 1 and 2 acting as a triode oscillator, and the cathode and grid 3 and the plate acting as a mixer, but in the transistor there are only the usual three elements to do everything.

At first you may wonder about this. However, there is no rule that says a transistor can only do one thing at a time. This one is a mixer and an oscillator simultaneously, just as if it were two entirely separate

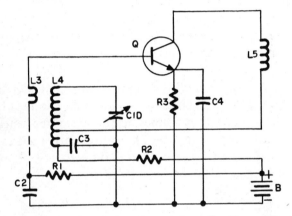

Figure 5.13 Transistor Converter—Oscillator Circuit

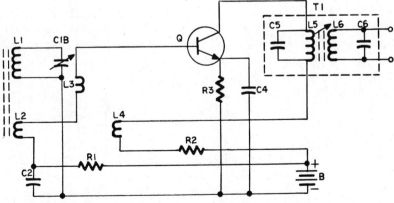

Figure 5.14 Transistor Converter—RF Amplifier Circuit

transistors. To make this clearer, Figure 5.12 has been redrawn in Figures 5.13 and 5.14 as a separate oscillator and mixer.

In Figure 5.13 you see a Hartley oscillator which only differs from that in Figure 4.2 by having the tap on L4 connected via L5 to the collector instead of to the emitter, and the oscillations of L4-C1D inductively coupled to the base by the "tickler" coil L3 instead of by a capacitor. These circuit variations do not change the principle of operation.

In Figure 5.14 you see that the mixer circuit is almost the same as that in Figure 5.10. If Figure 5.14 is superimposed on Figure 5.13 you get back Figure 5.12.

Circuit Variations

Pentagrid converters are pretty much standardized, but transistor circuits do offer some variety. An alternative to that shown in Figure 5.12 is shown in Figure 5.15.

As you can see, the main difference is that the "tickler" (L4) is now connected in the output circuit instead of in the input, while the oscillator resonant circuit is connected to the emitter instead of to the collector via the IF transformer. The net result is the same.

A connection for an optional external antenna is shown which is sometimes provided in the better type of receiver.

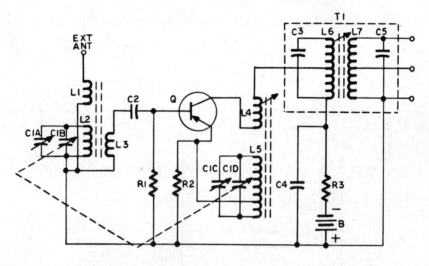

Figure 5.15 Transistor-Converter Circuit Variation

Modulation

The collector of the transistor is connected to a centertap on the IF transformer, which is often done, as explained in the section on IF transistor amplifiers in chapter 2.

TROUBLESHOOTING CHART FOR MIXERS AND CONVERTERS	SYMPTOM							
	TV – No Picture	TV – Snow in Picture	TV – High Band Missing	TV – Poor Contrast	TV or Radio – Interference	Radio – No Sound	Radio – Weak Signals	Radio – Background Noise
Tube or Transistor Faulty	X	X	X	X		X	X	
Wrong Aligment of Tuning Adjustment		X		X	X		X	X
Operating Voltage Incorrect	X	X	X	X		X	X	

6
DEMODULATION

Demodulators are the opposite of modulators. They recover the information from the modulated radio-frequency signal. For this reason they are also called detectors. Many demodulator circuits are built around one or more diodes, though triodes and specially designed tubes are used also.

In the process of modulation an RF and an audio or video signal were mixed together to produce a carrier with sidebands containing the audio or video information, which could then be radiated through space. At the receiver another signal was mixed with the RF signal and its sidebands to convert them to the IF frequency. Now, after amplification and removal of all undesired elements in the IF section, we have to get rid of the IF carrier in such a way that we are left with the audio or video information alone.

Detector circuits are also a source of feedback voltage to provide for automatic gain or volume control.

DIODE DEMODULATOR—CRYSTAL DETECTOR

Figure 6.1 shows a semiconductor diode used to demodulate an AM radio signal. Such a circuit may be called a diode demodulator or crystal detector circuit.

Demodulation

Distinguishing Features

This circuit contains a *semiconductor diode*, and is located between the last IF stage and the first AF stage of a superheterodyne AM receiver (radio or television).

The input is via the last IF transformer, and the output, in a radio receiver, usually is taken from the moving arm of a potentiometer (volume control).

In many cases a DC voltage for automatic volume (gain) control is extracted from the stage by means of a filter (R1, C2, R2, C3, in Figure 6.1).

No external DC is supplied to the circuit, for the diode is really a resistor which has much more resistance to current flowing in the direction of the arrow than against it.

Uses

AM radio and television receivers.

Detailed Analysis

In Figure 6.1 the output from the IF amplifier (see chapter 2) is coupled into L2, the secondary of T1, which with C4 is resonant at the IF frequency.

As this is an AC signal the upper and lower ends of L2 are alternately

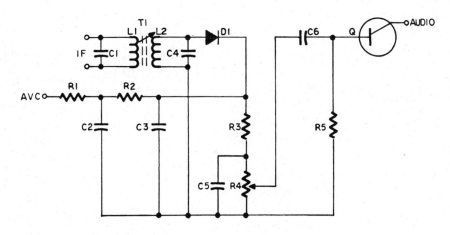

Figure 6.1   Crystal Diode Detector and AVC

positive and negative. If there were no D1 an alternating current would flow in the circuit connecting L2, R3 and R4, but as the diode only allows current to flow in the direction opposite to the arrow, it can only do so when the upper end of L2 is positive.

This results in a succession of pulses of direct current through R4 and R3 at the IF frequency, each one producing a voltage drop across R4 so that its upper end is positive with respect to its lower end. As a result C5 becomes charged so that its upper end is positive also.

During the interval between each pulse C5 starts to discharge by sending a current through R4 in the opposite direction, but before it can lose much the next positive pulse recharges it. As a result its charge stays very close to the amplitude of the positive voltage pulse developed across R4.

In the absence of modulation the IF carrier would be of constant amplitude, which would produce a constant voltage on C5. However, as you know, in amplitude modulation the amplitude of the signal is varying all the time in accordance with the audio information impressed on it, and so the voltage dropped across R4 varies, and the voltage on C5 rises and falls with it. The values of R4 and C5 are such that although C5 cannot discharge fast enough through R4 to follow the IF fluctuations it can follow the audio-frequency ones. The voltage across C5 and R4 therefore reflects the audio information originally impressed on the RF carrier. The movable arm of the potentiometer picks off as much of this AF voltage as will give the output desired, and couples it via C6 to the base of the audio amplifier transistor in the next stage.

Automatic Volume Control

The diode D1 is connected so that current pulses flowing through R4 and R3 give a positive voltage at the upper end of R3 with respect to the low side of the circuit. The filter R1, R2, C2, C3 smoothes out these pulses so that pure DC is provided for application to an earlier stage or stages of the circuit. This voltage varies only with the *average* amplitude, or strength, of the carrier, not with the audio fluctuations. For strong signals it will be higher, for weak it will be lower. When applied to previous stages it reduces the gain proportionately to overcome variations in the strength of the radio signal as received.

The polarity required for AVC voltage depends upon the application. For PNP transistors it will be positive, as shown, but for NPN transistors and vacuum tubes a negative voltage is necessary. In

Demodulation

this circuit reversing the polarity of the diode would reverse the polarity of the AVC voltage, if a negative voltage was required.

Circuit Variations

As explained above, the polarity of the diode may be reversed, or a different take-off point may be used to provide the AVC voltage of the proper polarity and amplitude.

When used in a TV receiver the diode detector circuit (Figure 6.2) has to handle the much greater bandwidth of the video signal (4.5 megahertz), while at the same time eliminating everything above that frequency. The values of C1, L3, C3 and L4 are chosen so that frequencies above the video frequency are so attenuated that they cannot get through the video amplifier to appear as interference on the TV screen. If a variable resistor like R4 (Figure 6.1) was provided in Figure 6.2 it would be a contrast control. However, this will more often be found in the next stage, the video amplifier discussed in chapter 2.

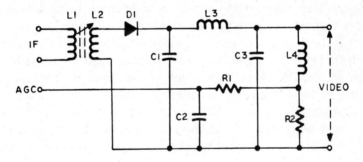

Figure 6.2 Video Detector

Automatic Gain Control

AGC considerations in the video detector are similar to those for AVC in a radio. They are different names for the same thing.

However, the type of AGC shown in Figure 6.2 is not always adequate for TV, and many receivers obtain it from a different circuit (Figure 6.4). Also, it is quite a common practice, if AGC is taken from the video detector, to make R2 a variable resistor, with R1 connected to the moving arm, so that the AGC voltage may be adjusted to suit local conditions.

Another circuit variation is the addition of a second diode as *AGC rectifier*, parallel with L4 and connected to conduct on negative alternations of the IF signal if negative AGC voltage is required (in which case D1 would face the other way also). In a vacuum-tube receiver a dual-diode tube may be used, or a triode connected so that the grid acts as the anode for one diode and the plate as the anode for the other, the cathode being common to both. Of course, the two diodes may also be semiconductors.

Peak Detector

As you have just seen, diode demodulators operate by *averaging* the amplitude of the rectified IF signal. However, if a single pulse at the IF frequency came along with a much greater amplitude it would pass unnoticed.

But if the circuit in Figure 6.2, for example, consisted only of the IF transformer, D1 and C1, then C1 would charge up to the maximum voltage of the pulse, and remain there because there would be no resistor through which it could discharge.

This is the principle of the *peak detector*, which is used for detecting *transients* (momentary voltage surges) that otherwise would not be observed. After each measurement of the voltage, the capacitor is discharged by a gate (see chapter 9), so as to reset it for further measurements.

The peak detector is *not* a demodulator, but is included here because it is a form of diode detector which you may run across, and therefore should know about.

VACUUM-TUBE DIODE DETECTOR

Vacuum-tube diodes perform in the same way as semiconductors, allowing current to flow through them in one direction only. They have the disadvantages of being larger and requiring a heater current to heat the cathode. However, the invention of the dual-diode-triode tube shown in Figure 6.3 enabled manufacturers to combine the diode detector function with the first audio amplifier stage in one tube, which was an important saving of cost in the heyday of vacuum-tube radios, and as these sets are still being produced the circuit is by no means superseded by the semiconductor type.

Demodulation

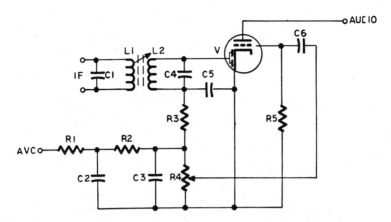

Figure 6.3 Vacuum-Tube Diode Detector and AVC

Distinguishing Features

The circuit contains a *dual-diode-triode tube*, and is located between the last IF stage and the audio output stage of a superheterodyne AM receiver.

The input is via the last IF transformer, and the output to the triode section of the tube is taken from the moving arm of a potentiometer (volume control).

An external DC supply is required for the operation of the triode section of the tube, but not for the diode. DC voltage for AVC is extracted from the diode stage by means of a filter (R1, C2, R2, C3).

Uses

AM tube radios.

Detailed Analysis

The input signal is coupled from the previous IF stage (see chapter 2) by the IF transformer L1-L2, which is tuned to resonance at the IF by C1 and C4. The triode section of V is used for audio voltage amplification, and as this circuit is explained in chapter 2 it will not be explained here. C6 and R5 in Figure 6.3 correspond to C1 and R1 in Figure 2.1.

The diode section of V consists of the cathode and two anodes beside it. When electrons are emitted from the cathode they will be attracted by a positive anode voltage, and current will flow. When the anode voltage is negative they will be repelled, and no current will flow. When the upper end of L2 is positive, current will flow in the circuit from the lower end of L2, via R3 and R4 to the cathode of V, and from the diode anodes back to the upper end of L2. When the upper end of L2 swings negative the diode blocks current flow in this circuit.

When the diode conducts, electrons flow from V down through L2 to charge C5 with a voltage equal to the amplitude of the IF signal. When the diode does not conduct, the charge on C5 begins to leak away through R3 and R4 (with which it is in parallel). However, if the values of C5 and R3 and R4 are chosen correctly C5 will not discharge very much before the next pulse recharges it, because it has to discharge through the high resistances of R3 and R4, whereas it charges through V, which has a very low resistance when conducting.

This means that C5 cannot follow the individual IF cycles, and remains virtually at the IF amplitude all the time. But this amplitude is fluctuating at the audio rate because of the audio signal with which it was modulated. The values of C5 and R3-R4 are such that the voltage on C5 can follow the audio fluctuations. Consequently the voltages across R3 and R4 vary with the audio modulation of the carrier.

The movable arm of the variable resistor R4 picks off as much of the voltage dropped across R4 as the user desires, and couples it through C6 to the next stage. C6 blocks the DC part of the signal, so that only AC appears at the grid of V.

The DC component for AVC is derived from the average or unmodulated amplitude of the IF carrier, and is extracted by the filter circuit R1, C2, R2, C3, as in the crystal diode circuit of Figure 6.1.

KEYED AGC

Simple AGC from the diode detector circuit, as shown in Figure 6.2, is not as efficient for TV as the circuit shown in Figure 6.4, because it cannot follow rapid fluctuations of the carrier strength, such as occur when an airplane flies over, when the picture would flutter severely. A keyed AGC circuit is designed to overcome this, and to reduce noise interference also.

Distinguishing Features

Although the circuit in Figure 6.4 looks like an amplifier you'll

Demodulation

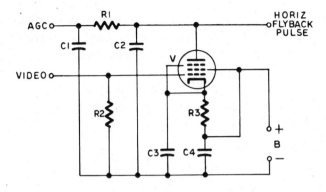

Figure 6.4 Keyed AGC

notice one startling difference right away. *The B+ voltage is connected to the cathode of the tube!* It is also connected to the screen grid, *but not to the plate* as in all regular amplifiers.

Instead, the plate is connected to the output of the horizontal deflection section, so as to apply powerful flyback pulses to it.

Uses

To provide AGC voltage in TV sets.

Detailed Analysis

DC Subcircuit: In Figure 6.4 B+ voltage is applied to the screen grid of V, and to the cathode via R3, but electron flow can only take place when a high-voltage pulse is present on the plate to make it positive with respect to the cathode. These pulses are frequently coupled to the plate through a capacitor to block DC voltages, and have a repetition rate of 15.75 kilohertz.

AC Input Subcircuit: A portion of the output of the video amplifier is applied to the control grid of V. The relationship between the cathode voltage, the control-grid voltage and the plate voltage (when a positive pulse is on it) is such that the grid voltage would prevent conduction at any time except when a horizontal sync pulse is received. In other words, the tube can only conduct when both a flyback pulse is present on the plate and a horizontal sync pulse is present on the control grid. Furthermore the amount of current it can pass depends upon the strength of the sync pulse on the control grid.

AC Output Circuit: The bursts of electrons flowing away from the plate charge C2 negatively. This capacitor acts like a storage battery, current flowing out of it between charging pulses through the filter network R1-C1.

The voltage on C2 varies with the amount of current passed by V, which in turn depends on the strength of the positive-going pulses applied to the control grid.

Circuit Variations

In the circuit in Figure 6.4 the plate of V would be connected directly to a separate insulated winding on the horizontal output transformer. In other cases the source of flyback pulses may be a tap on the main winding, the width control inductor, or damper circuit. As mentioned before, they are frequently coupled through a capacitor, and sometimes through a dropping resistor.

When flyback pulses are coupled through a capacitor, C2 may be replaced by a resistor. The coupling capacitor then acts as storage for the output current pulses from the tube.

The control-grid signal may be obtained from the sync amplifier instead of from the video amplifier.

A variable resistor is often placed in series with R3 and the B+ voltage in order to adjust the bias on the tube to compensate for the average strength of signals in the area where the set is operated. This is usually called the AGC control.

GRID-LEAK DETECTOR

Figure 6.5 illustrates a grid-leak detector. Though it is not often encountered anymore, it is particularly good on weak signals, so is sometimes used experimentally. However, some TV sets use it, as mentioned previously under automatic gain control.

Distinguishing Features

A triode with a tuned resonant circuit for RF or IF in its input, with an audio or video output.

The grid-leak resistor R1 may be shown paralleling C2, as in Figure 6.5, or connected directly to the low side of the circuit, as R1 in Figure

Demodulation

2.1. Effectively, the connections are the same, since the first really connects to the low side of the circuit through the coil.

Uses

Radio receivers (not superheterodyne), and some TV detector and AGC circuits.

Detailed Analysis

DC Subcircuit: In Figure 6.5 electron flow is from B– via V and R2 to B+. Electrons collecting on the grid of V return to the cathode via R1 and L2 (or R1 alone if connected direct to the low side of the circuit).

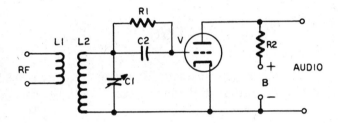

Figure 6.5   Grid Leak Detector

AC Input Subcircuit: RF or IF signals inductively coupled from L1 to L2 (which is tuned by C1 to be resonant at the desired frequency) are coupled by C2 to the grid of V. When the signal cycle is such that the upper end of L2 is positive, the lower end is negative. The grid and cathode of V act as a diode, and because the positive voltage of L2 is connected through C2 to the grid the "diode" conducts. When the signal polarity reverses so that the grid is driven negative, the "diode" does not conduct. This action is similar to the diode detectors discussed earlier in this chapter.

When the "diode" conducts, electrons flow from the cathode to the grid, and charge C2 so that the side connected to the grid is negative and the other side is positive. A voltage is developed across C2 which varies in amplitude with the modulation on the RF signal. Again, this action is the same as you saw in Figure 6.1, where an audio voltage was developed across C5.

As far as this audio voltage is concerned L2 is a straight piece of wire, so you have in effect only C2 and R1 connected between grid and cathode. V amplifies this audio voltage in the same way as any other audio voltage amplifier (see chapter 2).

AC Output Subcircuit: The output audio signal voltage appears across R2, and in a radio is coupled into the next stage, or to headphones. The use of this circuit in a TV set was covered under video detectors and AGC, and also as follows under circuit variations.

Circuit Variations

When this circuit is used in a TV video detector stage the cathode-grid diode replaces D1 in Figure 6.2. The triode amplifier output is then filtered and becomes an AGC voltage as in the AVC and AGC voltage circuits shown in Figures 6.1 through 6.4.

REGENERATIVE DETECTOR

The regenerative detector uses positive feedback to return a portion of the signal in the output to the input circuit as shown in Figure 6.6.

Distinguishing Features

The feedback coil L2, the RF input, the audio output and the variable resistor R1 *in the grid circuit* of V, taken together, enable you to tell this circuit from an RF amplifier or an oscillator.

Uses

Sometimes used in experimental receivers because it is very sensitive, and can be used to receive both modulated and CW signals.

Detailed Analysis

DC Subcircuit: In Figure 6.6 electron flow is from B– via V, L2 and the primary of T1 to B+.

AC Input Subcircuit: The RF signal picked up by the antenna is coupled from L1 to L3, which is tuned to resonance by C1, and then applied via C2 to the diode consisting of the grid and cathode of V.

Demodulation

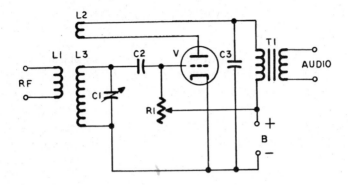

Figure 6.6   Regenerative Detector

Demodulation takes place in the same way as in the grid-leak detector of Figure 6.5.

AC Output Subcircuit: The audio and RF signals present on the grid appear in the output after amplification by the triode V. The audio is coupled to headphones or to another stage. The RF signal is coupled by L2 back into the input, where it reinforces the signal applied to the grid. The degree of coupling sometimes can be adjusted by moving L2 closer to or further from L3. However, it is more usual to adjust the bias on the grid of V by means of R1. The return path for RF is by means of C3 which bypasses T1 for RF, but offers a high reactance to audio signals, and does not bypass them.

Adjusting R1 increases the sensitivity of the circuit by increasing the gain of V, until the amount of feedback becomes sufficient to cause oscillation. The most sensitive point for reception is just below the point where oscillation begins.

However, when listening to CW (code) the signal produced when oscillating beats with the incoming signal to give an audio difference frequency which can be heard in the headphones or speaker.

Circuit Variations

A transistor may be used instead of a tube, and the audio output transformer is not required with headphones. As mentioned previously, L2 (which is sometimes called a "tickler" coil) may be movable to adjust the degree of coupling, in which case R1 may be replaced with a grid-leak resistor.

FM DISCRIMINATOR

With the schematic illustrated in Figure 6.7, we begin discussion of a group of circuits used in FM demodulation. The FM discriminator is not used much anymore because it requires a *limiter* (or noise suppression) preceding stage, whereas the *ratio detector* discussed next does not. But you may still meet with it (for example, one was shown in block form in the section on FM modulation in the previous chapter, in Figure 5.3).

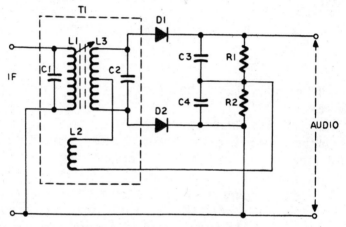

Figure 6.7   FM Discriminator

Distinguishing Features

An IF transformer with an extra winding connected to a centertap on its secondary winding.

The opposite ends of the IF secondary connected to two diodes, *both facing the same way.*

Uses

FM radio receivers, TV sound sections and FM transmitters.

Detailed Analysis

L3 and C2 form a resonant circuit tuned to the intermediate frequency, which is 10.7 megahertz in an FM radio or 4.5 megahertz in

Demodulation

TV. As you know, a resonant circuit seems like a pure resistance to a signal as its resonant frequency; that is to say, the current and voltage are in phase. But when the signal frequency is at some other frequency the resonant circuit acts like an inductor or a capacitor, depending upon whether the frequency is higher or lower.

T1 is constructed so that when the signal induced in L3 is exactly at the resonant frequency it will be 90 degrees different in phase from the signal at its resonant frequency; that is to say, the current and voltage winding it is in phase.

The signal voltage in L2 combines with the signal voltage in each half of L3 to produce voltages at each end of L2 that are equal and opposite. When the upper end of L3 swings negative the lower end swings positive, and vice-versa.

When the frequency changes, due to modulation, the phase angles between the signal voltages in L3 and the signal voltage in L2 change accordingly, but in opposite directions. This results in unequal voltages at the opposite ends of L3.

When the upper end of L3 is positive, electrons flow from the upper side of C3, and flow to L3 through D1. At the same time the lower end of L3 is negative, but no current can flow through D2. On the next half-cycle, when the voltage on the lower end of L3 swings in a positive direction, C4 becomes positively charged also.

The charging current consists of a train of pulses that, as we've just seen, charge C3 and C4 positively. Between pulses these capacitors do not have time to discharge through R1 and R2, since these resistors have a high value, so each becomes charged with the average peak voltage across the corresponding half of L3.

As long as the voltages on the opposite ends of L3 are equal the charges on C3 and C4 are equal. Likewise, the voltages at the two audio output terminals are equal, and therefore cancel each other out, so no output signal is produced.

But when the voltages on C3 and C4 differ, because of frequency modulation, unequal voltages appear at the output terminals. These alternate at the audio rate, because while the time-constants of C3-R1 and C4-R2 are such that they cannot follow the IF fluctuations in the signal voltage, they are designed to keep up with the audio modulation rate.

Circuit Variations

Some discriminators do not have a third winding on the IF

transformer (L2). The centertap is connected to the upper end of L1 by means of a capacitor. A dual-diode vacuum tube may also be used instead of two semiconductors.

RATIO DETECTOR

There is a close similarity between the ratio detector (Figure 6.8) and the discriminator (Figure 6.7).

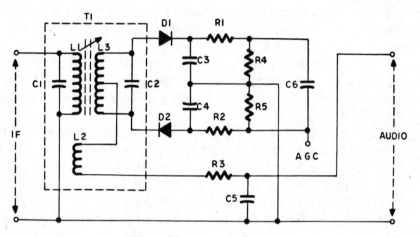

Figure 6.8 Ratio Detector

Distinguishing Features

An IF transformer with an extra winding connected to a centertap on its secondary winding.

The opposite ends of the secondary are connected to two diodes, *facing in opposite directions.*

Uses

FM radio receivers and TV sound sections.

Detailed Analysis

As already mentioned, the ratio detector has replaced the

Demodulation

discriminator because it does not require a preceding limiter (clipper) stage, so is more economical to produce.

The operation of the transformer T1 is the same, however, and need not be repeated here. But because D2's polarity is reversed the voltages appearing on C3 and C4 are now in series. A large capacitor C6 (usually an electrolytic of 5 microfarads) is connected in parallel with both C3 and C4. Because of its large capacitance it damps out noise coming in with the signal.

The total voltage across C3 and C4 will equal the voltage across C6, which is the same as the total voltage across L3. However, as the signal swings back and forth in frequency, the *ratio* of the voltages on C3 and C4 will change with respect to each other. Consequently, the voltage at the point between them will fluctuate accordingly. A voltage difference varying at the audio rate will also exist between this point, which is connected to one side of the audio output.

Circuit Variations

The lower end of L2 may be connected to the junction of C3 and C4, and the junction of R3 and R4 will then be omitted. The audio output signal will then be taken from these two junctions.

C3 and C4 may be omitted altogether if circuit capacitances are adequate.

C3, C4, R1, R2, R3 and R4 may be encapsulated together in an integrated circuit (IC). In the schematic they will be enclosed in dashed lines to indicate this.

GATED-BEAM FM DETECTOR

Figure 6.9 illustrates a typical gated-beam demodulator, the most economical FM detector, since it combines limiting, detection and audio amplification in one stage.

Distinguishing Features

A vacuum tube of special construction is used in this circuit, such as a 6BN6, 6JC6 or 6HZ6. In the schematic it has the same symbol as a pentode. Its three grids, reading from nearest the cathode, are named and connected as follows:

1. *Limiter grid:* IF input.
2. *Accelerator grid:* B+ voltage.
3. *Quadrature grid:* parallel resonant circuit with quadrature coil.

Plate and cathode connections are as in other amplifiers.

Uses

FM radio or TV sound demodulation stage.

Detailed Analysis

DC Subcircuit: In Figure 6.9 electron flow is from B– via R1 to V. From V's plate it returns via R2 and R5 to B+. From the accelerator grid, current flows to B+ via R6.

AC Input Subcircuit: The IF signal is coupled from the previous stage by T1 to the limiter grid of V. The construction of this tube is such that it has a very steep characteristic curve. A swing of two or three volts is enough to drive it from saturation to cutoff (see Appendix for explanation of tube operation). The value of R1 is chosen so that the grid bias is about –4 volts, or midway on its characteristic curve. Consequently the action of the grid is to clip or limit the signal so that it has a constant amplitude, and noise and interference are removed. The action of this grid is to switch the electron current from the cathode on and off in a series of flat-topped, steep-sided pulses, at the frequency of the IF signal. The electrons of these pulses are focused in a beam and accelerated by the second grid, which has a voltage of around 100 volts.

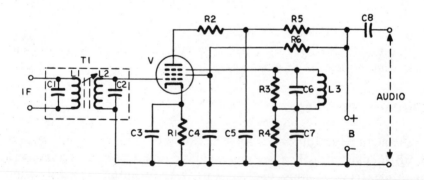

Figure 6.9 Gated-Beam FM Detector

Demodulation

Quadrature Grid Subcircuit: Connected to the third grid is a resonant circuit consisting of C6 and L3. R3 and R4 are not really part of this circuit as their values are too high to have any loading effect on the circuit. They provide a path to ground for electrons collecting on the quadrature grid. (In some circuits they are omitted.)

This resonant circuit is connected between the quadrature grid and the cathode (by way of C7 and C3). It is tuned to the IF resting frequency. In the absence of modulation a stream of electrons travels from the cathode to the plate in pulses, as previously explained, and some of the electrons in each pulse strike the quadrature grid, causing pulses of current to flow in the resonant circuit. As this circuit is tuned to their frequency it oscillates, causing the voltage on the quadrature grid to swing back and forth. When it swings negatively it cuts off the flow of electrons to the plate, but when positive it allows them to pass; so, like the limiter grid it also acts as a switch.

It is switching at the IF resting frequency, and as long as the limiter grid (in the absence of modulation) is doing the same the pulses of electrons passed by the two grids are at the same rate. However, the construction of the tube is such that the positive swings of the quadrature grid lag behind those of the limiter grid by 90 degrees.

It is like having two gates opening and closing in such a way that the second gate opens half way between the opening and closing of the first, and then closes half way between the closing and opening of the first. Electrons can only flow when both gates are open, which is during the second half of each positive swing of the voltage on the limiter grid.

However, changes in frequency of the input signal caused by modulation change the amount by which the opening and closing of the "gates" overlap. The greater the overlap, the longer the burst of current, and vice-versa.

AC Output Subcircuit: These bursts of electrons arrive at the plate, and thence flow through R2 to charge C5. This RC combination is what is called an *integrating circuit* (see chapter 8). Wider bursts mean more electrons, causing the voltage on C5 to build up; narrower bursts allow it to fall. Since the width of the bursts varies with the frequency, which in turn varies with the amplitude of the modulating signal, you can see that the voltage on C5 will rise and fall proportionately to the audio voltage with which the carrier was frequency modulated.

The tube is basically a voltage amplifier (see chapter 2) even though it also performs the limiting and demodulating functions just described. Consequently the output is usually sufficient to drive an audio output stage, whereas the ratio detector requires an audio amplifier to follow it and drive the output stage. The discriminator not only requires an audio voltage amplifier to follow, but a limiter to precede it. (Limiters, or clippers, are covered in chapter 8.) Since the gated-beam tube enables limiting, demodulation and voltage amplification to be done in one stage it is easy to see why this is such a popular circuit in vacuum-tube FM receivers.

Circuit Variations

As mentioned before, R3 and R4 may be omitted, and the lower end of L3 may be connected directly to the low side of the circuit instead of via C7. Another variation sometimes met is the replacement of R1 with a variable resistor called a *buzz control*. This adjustment is provided to select the best bias point for the limiting grid, which is that which gives the least hissing sound with a weak signal. The quadrature coil often has an adjustable slug which may also be adjusted for the strongest and clearest sound. The cathode bypass capacitor C3 is also omitted sometimes for the reasons explained in chapter 2 under *degeneration*.

The output signal is usually coupled through C8 to a volume control, as in other detectors.

ENVELOPE DETECTOR

Figure 6.10 shows a switching-bridge detector, the most common form of envelope detector. Envelope detectors are named so because they are used to detect the *envelope* of the complex signal used in stereo FM broadcasting.

Distinguishing Features

A four-diode bridge follows an IF transformer, and has left- and right-channel outputs.

The detector is the final stage of a multiplex circuit, the rest of which is illustrated in block form in Figure 6.10. Each of the other circuits is discussed elsewhere in the book.

Uses

Envelope detectors are used in stereo-multiplex circuits to separate the right and left channels of a stereo broadcast.

Demodulation

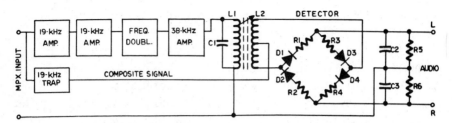

Figure 6.10   Envelope Detector in Stereo Multiplex Circuit

Detailed Analysis

FM stereo broadcasts must be receivable by monaural equipment also, therefore the audio from both channels is mixed together at the station in what is known as the L + R signal. Thus a single-channel set receives both channels of a stereo transmission. This is what is meant by the expression "compatible FM stereo."

To provide for two-channel reception an additional audio signal is generated at the transmitter by splitting the two channels, so that there are *two* right- and *two* left-channel signals. Then the phase of the second right-channel signal is inverted before mixing it with the second left-channel signal. There are now an L + R and an L − R signal, the L − R representing the *difference* between the two channels.

Both signals are used to modulate the carrier, but in order to keep them separated the L − R signal is first made to modulate a 38-kilohertz subcarrier, using AM. Sidebands are produced above and below this frequency over the range from 23 kilohertz to 53 kilohertz.

This 38-kilohertz signal is then used to modulate the carrier along with the L + R audio signal. However, so as to improve the signal-to-noise ratio, the 38-kilohertz carrier is suppressed; consequently the signal received by the FM receiver consists of the FM carrier, modulated by the audio and the 38-kilohertz sidebands only.

In order to recover the L − R signal it is necessary to re-insert the 38-kilohertz subcarrier, which is done in the multiplex circuit of the receiver, or "stereo adapter" as it was called when purchased separately. A locally generated signal by itself will not do, because it must have exactly the same frequency and phase as the original. Therefore a pilot signal of 19 kilohertz is also transmitted. This doesn't affect the signal-to-noise ratio, because it is transmitted in the unused space between the maximum audio (L + R) frequency of 15 kilohertz and the lower limit of the lower L − R sideband, which as you already saw is 23 kilohertz. This pilot is also of lower amplitude than the subcarrier before suppression.

The output of the FM detector—probably a ratio detector—contains the L + R audio signal, the L − R 38-kilohertz sidebands and the 19-kilohertz pilot. In a monaural receiver the 19-kilohertz and 38-kilohertz signal cannot pass through the audio output stages to the speaker, therefore only the L + R audio would be heard.

But in a stereo set the composite audio output of the ratio detector will be routed through a stereo-multiplex circuit similar to that in Figure 6.10, where the 19-kilohertz pilot signal is extracted and amplified by IF amplifier stages similar to those described in chapter 2. It is then applied to a frequency-doubler, which reconstitutes it as the original 38-kilohertz carrier. (Frequency-doubling is discussed in chapter 12.)

Meanwhile the composite audio signal is fed to the secondary of the 38-kilohertz IF transformer in such a way that the L − R sidebands are recombined with their carrier.

The diodes in the detector bridge are arranged so that D1 and D3 demodulate the 38-kilohertz signal when it swings so that the upper end of L2 is negative, and D2 and D4 demodulate it when the lower end of L2 is negative. This gives demodulated audio from D1 and D3 which is L − R, while that from D2 and D4 is of opposite phase, or −L + R. These combine algebraically with the L + R signal, which is present also, of course, as follows:

$$\begin{array}{cc} L + R & L + R \\ \underline{L - R} & \underline{-L + R} \\ 2L & 2R \end{array}$$

The left and right channels then pass through separate audio amplifiers to the left- and right-channel speakers.

Circuit Variations

The circuit we have just discussed uses four diodes in what is called a switching bridge because it switches back and forth as the 38-kilohertz carrier swings back and forth between its positive and negative peaks. In other words, it *samples* each channel alternately. This can also be done with only two diodes, operating alternately to charge capacitors as in the discriminator, except that the audio signals appearing on the capacitors are now for separate channels.

A rather more complex circuit, using bandpass filters (see chapter 8), separates the output of the FM detector into its three components: the L + R signal (up to 15 kilohertz); the 19-kilohertz pilot signal; and the L − R signal (23 to 53 kilohertz). In this circuit, which is called a *bandpass and matrix* circuit, a 19-kilohertz local oscillator generates a signal which is synchronized by the 19-kilohertz pilot (as explained in chapter 4). This is then doubled and re-inserted to reconstitute the original L − R signal. This signal is then demodulated, using two diodes as just described, to give the L − R and − L + R signals, which are then combined with the L + R signal in an arrangement of capacitors and resistors called a matrix, which is to ensure that they are combined in the right proportions.

DE-EMPHASIS

In chapter 5 you saw that in FM transmission it was necessary to use *pre-emphasis* to obtain a more favorable signal-to-noise ratio. This resulted in the higher audio frequencies being transmitted with greater power than the lower. In the receiver it is necessary to restore the proper proportion between the higher and lower frequencies, and this is done by *de-emphasis*.

In Figure 6.8 de-emphasis is performed by R3 and C5, which have a time-constant corresponding to that of the pre-emphasis circuit in the transmitter. Such a high-frequency attenuation circuit is similar in operation to the bass control of chapter 2, but without the variable resistor.

While de-emphasis must take place between the detector and the audio amplifier following it, it cannot come before a stereo-multiplex circuit, since it would attenuate the 19-kilohertz pilot and the 38-kilohertz sidebands. Consequently it will always be placed in the output of a multiplex unit. However, sets designed to receive stereo broadcasts usually have a function switch with a monaural position which bypasses the multiplexer when not required. In this case the FM detector has two outputs, one for stereo without de-emphasis, and one for monaural with de-emphasis, and the switch selects the appropriate output.

SYNCHRONOUS DETECTOR

In the previous chapter you saw how phase modulation as used in color TV was accomplished by using a pair of doubly balanced modulators. PM demodulation is the same process in reverse. Figure 6.11 shows a synchronous detector using a pentode. Two of these are required to extract the signal's two component vectors, just as two modulators were required to combine them.

Distinguishing Features

The 3.58-megahertz reference signal is applied to the No. 3 grid of the pentode, and the modulated color subcarrier (chroma) is applied to the control grid. In the output appears a filter circuit (L1-C1) to eliminate the 3.58-megahertz subcarrier, which is no longer required.

Uses

Color television reception.

Detailed Analysis

DC Subcircuit: In Figure 6.11 electron flow is from B– via R2 to V1. From the plate of V1 current returns via L1, R4 and L2 to B+. From V1's screen grid it reaches B+ through R3.

AC Input Subcircuit: The modulated subcarrier (chroma) is applied across R1, *the color gain control*, which is adjusted to give the desired signal on the control grid. This is biased to avoid complete cutoff, so that the whole color signal is amplified.

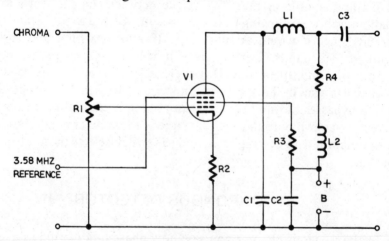

Figure 6.11 Synchronous Detector

Demodulation

The 3.58-megahertz reference signal, synchronized with the transmitter, is applied to the No. 3 (suppressor) grid. A negative voltage on this grid will cut the tube off completely. In a receiver designed for I and Q color signals the reference signal for the I demodulator must be phase-shifted 57° before reaching the grid. That for the Q signal must be 90° more, as you saw in the previous chapter. However, many receivers use different vectors for which other phase angles are required. This does not alter the principle of operation of this circuit.

AC Output Subcircuit: V1 can only conduct when the suppressor grid is not negative. During each positive half-cycle of the 3.58-megahertz reference signal, current will flow in proportion to the voltage on the control grid. The greatest current will flow when a positive half-cycle of the color signal on the control grid coincides with a positive half-cycle of the reference signal on the suppressor grid. This happens when both signals are in phase.

However, the phase angle of the color signal is constantly changing as the color information changes. Therefore the direct current out of V1 will consist of a series of pulses at 3.58 megahertz which will vary in width and amplitude according to the amount of overlap between the two signals. This action is similar to that of the gated-beam tube previously discussed in this chapter.

Since the reference signal of each of these demodulators is 90° different from the other, their output will consist of the two vectors added at the transmitter. In each demodulator these pulses are integrated and smoothed by C1 and L1, and coupled through C3 to the next circuit.

Circuit Variations

Demodulators sometimes employ two gated-beam tubes of the type used in FM detection instead of pentodes. Dual-beam-deflection tubes have been developed (such as the 6JH8) to give a push-pull action.

Transistors can replace tubes in essentially similar circuits. Diodes may also be used in circuits resembling the ratio detector already discussed (see Figure 6.8). Two pairs of diodes are required, but they share a centertapped transformer that supplies reference signals at 90° phase difference. The diodes conduct according to the phase and amplitude of the color signal.

TROUBLESHOOTING CHART FOR DEMODULATORS	SYMPTOM							
	Radio — Sound Absent or Weak	Radio — Sound Intermittent	Radio — Hum	Radio — Buzz or Hiss	Radio — Distortion	TV — No Picture	TV — Contrast Poor	TV — Interference Pattern, or Grainy
Tube Weak, Dead	X		X		X	X	X	
Tube Socket Contacts Dirty or Broken		X						
Volume Control Dirty or Defective	X	X						
IF Transformer Misadjusted	X			X	X			
Buzz Control Misadjusted	X			X				
Quadrature Coil Misadjusted	X			X	X			
Cathode-Bypass Capacitor Defective			X	X	X			
Storage Capacitor (Ratio Detector)				X				
De-emphasis Filter					X			
Peaking Coil (L3 in Figure 6.2)							X	
Peaking Coil (L4 in Figure 6.2)							X	X

7
POWER SUPPLIES

A power supply is a circuit which provides the operating voltages for the other circuits in a piece of electronic equipment. Except for heater current for tubes, all operating voltages are DC. While small radios can obtain all their requirements from batteries, most electronic equipment plugs into the 117-volt power line. This does not mean that all power supplies convert AC to DC, but the majority do. In their circuitry they are similar to the AM demodulation circuits described in chapter 6. There are also independent power supplies, portable or otherwise, which are not part of some other piece of equipment.

BATTERY POWER SUPPLY

The simplest and one of the most widely used sources of power for electronic equipment is a *battery*. The same symbol is used, regardless of type. The long line is always positive, but polarity may be indicated in addition, as in Figure 7.1. Sometimes a distinction is made between single cell and multicell batteries, symbolizing the first by a single pair of lines, one short and one long, although this may be used also to indicate only a generalized direct-current source. Voltage value may also be shown.

Switches are usually shown as S1 or S2, in Figure 7.1. They are usually shown open.

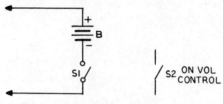

Figure 7.1   Battery Power Supply

Battery power supplies seldom have any other components, unless they are rechargeable-battery supplies. These have a simple rectifier-type supply as, for example, that illustrated in Figure 7.7. (L1 will probably be a resistor, however.) On most radio and TV receivers the physical switch is mounted on the rear of the volume control, and is turned on or off by the same knob. This may be indicated in a schematic by a statement beside the switch symbol (as by S2 in Figure 7.1), or by a dashed line connecting the switch symbol to the variable resistor symbol to indicate they are ganged together. In this way the two symbols can be located according to their functions.

UNiVERSAL AC-DC POWER SUPPLY

This power supply got its name because it could be operated from both AC and DC power lines, as it does not use a transformer. You are not likely to encounter a DC power line today, but the name is supposed to sound better than "transformerless power supply." Figure 7.2 gives a typical example.

Distinguishing Features

1. Rectifier (diode) tube, or solid-state rectifier.
2. Pi-filter, in which R1 may be replaced by a choke.
3. Series-string of filaments across the power line, as shown.
4. No power transformer.
5. Connection to power line, and on-off switch.

Detailed Analysis

In this supply the tube heaters are connected in series across the power line, which is assumed to be at 120 volts. As you know, the first

Power Supplies

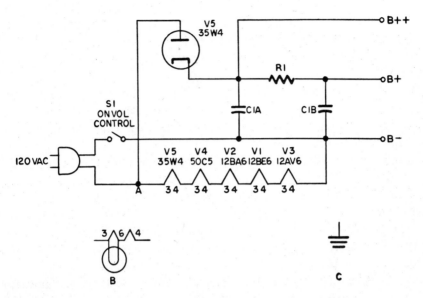

Figure 7.2   Universal AC-DC Power Supply

figure, or figures, of tube numbers give the heater voltage required. Those shown in the diagram add up to 121 volts.

Five tubes, each requiring 24 volts, would also add up to 120 volts. However, when this circuit was adopted heater voltages had already been pretty well standardized at 6.3 or 12.6 volts, which were the battery voltages originally used. Furthermore, using higher AC voltages for heaters introduces hum when the heater current is AC. Consequently the lower-voltage tubes already available were used in the circuits most susceptible to hum pickup, and two special tubes were developed for the less sensitive audio output and power supply circuits. By arranging them so that the most sensitive circuits' tubes are also at the low end of the string (lowest AC voltages), hum pickup has been kept below an audible level.

The numbers 3 and 4 indicate the tube pins for the heaters. By showing the heaters separately in the power supply like this, it saves showing them in the tube symbols, and avoids the additional congestion and confusion of having to draw heater connections running all over the schematic.

The AC of the power line is converted to DC by the rectifier tube V5. When the side of the power line connected to the plate swings positive the other side swings negative and supplies electrons via S1 to the low side of the circuit, B-. Current flows from B- to the cathodes of the four tubes in the radio, from the cathodes to the plates and screen grids, and back to B+ and B++. From these points it reaches the cathode of V5, flows to the plate, and from there to the positive side of the power line.

When the AC line reverses polarity on the alternate half-cycle, V5's plate becomes negative with respect to the cathode, and V5 cannot conduct. Consequently, bursts of current flow through V5 in one direction only on the half-cycles that make its plate positive, but no current flows on the alternate half-cycles; hence it is called a *half-wave rectifier*, which means that half the power available from the power line is not being used. For the small power requirements of this type of radio this is not important.

These pulses of DC cannot be used for most of the tubes, as they would still create a loud hum, so a pi-filter circuit is provided to smooth them out and give a constant voltage at B+. The filter consists of C1A and C1B, which are two sections of a dual electrolytic capacitor (usually from 30 to 50 microfarads each), and R1, a resistor capable of handling the plate supply current of V1, V2 and V3. Its value will probably be around 1000 ohms.

Filters are covered more fully in chapter 8. Without going into too much theory we can say that in this one, when V5 conducts, C1A charges up to the voltage at V5's plate, since the tube has a low resistance when conducting. When V5 cuts off it cannot conduct, so C1A starts to discharge through R1 and the rest of the circuit. The circuit resistance is much higher, of course, so C1A has not discharged very much before V5 starts conducting again on the next positive half-cycle and replaces whatever was lost. The voltage on C1A therefore fluctuates between the maximum value and that to which it falls when V5 is not conducting.

This fluctuating voltage can be thought of as consisting of two components: a steady DC voltage at the lower value, and an AC voltage ripple "riding" on it. This AC component is still too much for our B+, so R1 and C1B are provided as a voltage divider to reduce it to an acceptable figure. For AC at 60 hertz, C1B has a reactance of 100 ohms or less, whereas R1 is at 1000 ohms (or thereabouts). This divides the AC voltage appearing between R1 and C1B by a factor of at least eleven. However, as C1B offers a very high resistance to DC the latter is hardly affected at all. For small loads this amount of filtering is quite adequate.

The one tube that draws a heavier plate current than the rest (V4) is supplied directly from C1A (B++ supply). As this tube is an audio

Power Supplies

power amplifier, fluctuations in the plate *voltage* have very little effect on it (see chapter 3), and the ripple on its supply is therefore of no consequence.

Circuit Variations

As you just saw, resistor R1 is perfectly adequate as part of the pi-filter as long as the amount of current required is small (and when the plate supply of V4 is drawn directly from C1A). Therefore, in small and inexpensive clock radios and the like this will be the normal arrangement. However, as soon as a higher current requirement arises, as in a set with more tubes, or with pretensions to high fidelity, then a choke will be substituted for R1. A 5-henry choke presents a reactance of approximately 2000 ohms to the AC component, while offering negligible resistance to the DC component. As a result higher DC currents can be passed without dropping the voltage. For AC the effect is the same as when a resistor is used. You might meet this type of filter more in TV sets today, as we shall see later on in this chapter.

A pilot light is sometimes connected across half of V5's heater filament, as shown at B. This not only tells you that the set is on, but also acts as a safety device. The plate of V5 is connected to the centertap of the filament (pin 6) instead of to point A (in the main diagram). If a short develops somewhere in the set so that a heavy current is drawn through V5, the half of the filament between pins 3 and 6, and the pilot light, will burn out as if they were a fuse.

A ground symbol (as at C) is frequently drawn at B– to indicate a chassis ground return. This saves drawing in the low side of the circuit. The same symbol is then used at other points in the set to indicate points also connected to the chassis, and therefore to B–. This has become such a habit with radio and television manufacturers and others that they do it even though there may be no real chassis (as in today's printed-circuit boards). It has a distinct advantage in small crowded diagrams, but in the much larger prints used in industrial work it is more usual to draw in the "ground return" or low side of the circuit.

FULL-WAVE RECTIFIER

A disadvantage of the AC-DC power supply was that drawing a heavy current from it would cause the voltage on C1A to fall excessively between the half-cycles when V5 conducted. While this could be overcome by elaborate filtering it is more economical to charge the input filter capacitor on both half-cycles. Full-wave rectifiers which do this are shown in Figures 7.3 and 7.4.

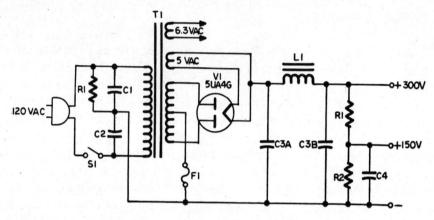

Figure 7.3   Vacuum-Tube Full-Wave Power Supply

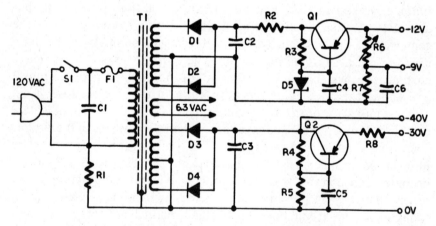

Figure 7.4   Solid-State Full-Wave Power Supplies

Distinguishing Features

This type of power supply uses a power transformer with various windings. Diodes are connected to the opposite ends of the secondary winding or windings used for the DC supply. In vacuum-tube types the

Power Supplies

diodes may be in one tube (as in Figure 7.3), or in separate tubes (more usual for higher voltage supplies).

Input connections from the power line are indicated by illustrating a plug or similar connection, switch, fuse, line filter, all on the primary side of the transformer.

The output is via a filter and voltage divider network. There may be more than one supply, as in Figure 7.4.

Uses

TV, radio and other electronic equipment with comparable requirements.

Detailed Analysis

Vacuum-Tube Full-Wave Power Supply: The input to this power supply is shown as a simple two-wire cord and plug. However, a third wire could be used, and would be connected to the junction between C1 and C2. These capacitors are usually of .01-microfarad capacitance and form, with R1 (100 kilohms), a filter for reduction of power-line interference (pops from operation of switches, interference from motors, and so on). A *switch* (mounted as part of the volume control), and sometimes a *fuse*, will also be part of the input circuit.

The principal secondary winding is the high-voltage winding, of which the opposite ends are connected to the two plates of V1. The centertap is connected to the low side of the circuit via the fuse F1. This centertap is the most negative point in the circuit.

The filament of V1 is heated by a special 5-volt AC winding. Electrons flow from this filament alternately to each plate according to which end of the winding is positive. The resultant DC current is drawn through the filter and voltage divider from all the tubes in the set, which in turn draw electrons from their cathodes, which are supplied from the low side of the supply.

The pi-filter (C3A, C3B, L1) operates in the same way as discussed under the AC-DC supply, and the voltage divider divides the +300V supply equally to give a +150V supply for the more sensitive tubes. C4 gives additional filtering to this supply.

Tube heaters get a 6.3 AC voltage from the other filament winding. Some power transformers have more than one 6.3V winding, or a 12.6V winding with a centertap, or both. Color-coding for leads for power transformers is given in the Appendix.

SOLID-STATE FULL-WAVE POWER SUPPLY

Figure 7.4 really depicts two power supplies sharing one power transformer. The remarks made about the input to Figure 7.3 power supply also apply to this one, but the two main secondaries give rise to two sets of voltage outputs. The pair of semiconductor diodes in each case function in the same way as the 5UA4G tube of Figure 7.3, but require no heater current, so no filament winding is provided. The 6.3-volt winding shown is for dial lamps and such.

Transistors Q1 and Q2 are provided for voltage regulation, discussed later in this chapter.

Circuit Variations

The circuits of Figures 7.3 and 7.4 appear in many variations. Sometimes there are two rectifier tubes instead of a dual tube, sometimes two dual tubes are connected in parallel. As already mentioned, there can also be differences in the number and arrangement of windings on a power transformer.

Line filters can vary also from the examples shown, though the principles are the same.

The filter, voltage divider and regulation circuits can sometimes get quite complicated where many different supply voltages are required. However, they all work according to the principles explained in this chapter.

TRANSFORMERLESS VOLTAGE-DOUBLER POWER SUPPLY

Figure 7.5 illustrates a very popular type of power supply. It is similar to the AC-DC power supply of Figure 7.2, but capable of higher voltage and current output, which makes it suitable for economy TV use.

Distinguishing Features

1. Two diodes (rectifiers), usually solid-state.
2. Pi-filter with choke.
3. Series-string of tube filaments.

Power Supplies

4. No power transformer.
5. Connection to power line, on-off switch, fuse or fusible resistor.

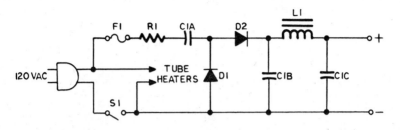

Figure 7.5 Transformerless Voltage-Doubler Power Supply

Uses

Radio, TV and the like.

Detailed Analysis

With S1 closed the alternating power-line voltage is applied across C1A and D1. The lower side of D1 is alternately positive and negative. When positive, D1 conducts, allowing electrons to leave the right-hand side of C1A. The other side of C1A is charged negatively. As D1 can only conduct in one direction the positive charge on C1A builds up to the peak value of the power-line voltage (120 × 1.4 = 168V). This positive voltage stored on C1A acts as a base, with the power-line AC superimposed on it, so that the voltage at the upper end of D1 is the algebraic sum of the DC and AC peak voltages. This voltage varies between +336 and 0 (+168 ±168) volts. The action of the rest of the circuit to the right of D1 is the same as that of Figure 7.2, except that the filtered DC voltage will be approximately 250 to 300 volts instead of 100 to 150 volts.

R1 is a *surge resistor*, a low-voltage resistor which controls the rush of current into C1A when S1 is closed, which might otherwise destroy D1. It can be combined with the fuse F1 in a *fusible resistor* ("surgistor"). This can withstand the heavy charging current which only lasts for a very short time, but melts if a short elsewhere in the set draws a sustained heavy current.

Circuit Variations

Although fundamental principles remain the same there are several different ways of arranging these circuits. Much of the variation is in voltage dividers and filters (discussed later).

A disadvantage of this type of circuit is that one side of the chassis (when using a metal chassis) is connected to the power line. To avoid this the low side may be connected to the chassis through a large capacitor located in series with C1B. This capacitor then replaces C1A, which is not required.

Considerable variation is also possible in the order in which the series-string of tube heaters is arranged. If there are very many it may be necessary to arrange them in two parallel strings (see Figure 7.8). Occasionally a filament transformer is used instead, and the tubes are then connected in parallel. A filament transformer is a less expensive type of transformer which reduces the power-line voltage to 6.3 or 12.6 volts. (Such transformers are used for domestic purposes also, such as bells, thermostats, and so on.)

TRANSFORMERLESS VOLTAGE-TRIPLER POWER SUPPLY

The voltage-tripler circuit illustrated in Figure 7.6 works in exactly the same way as the voltage-doubler of Figure 7.5. Compare the two circuits. (Switch, fuse, heater connections and surge resistor have been omitted for convenience.)

C1 and D1 store a positive charge in the same way as C1A and D1 in Figure 7.5. This additional voltage brings the output up to between 400 and 450 volts. Further capacitor diode pairs could be added to make voltage-quadruplers, -quintuplers, and so on, but capacitors that can

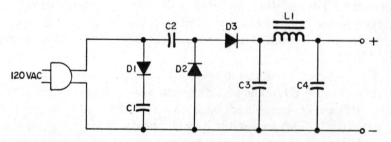

Figure 7.6 Transformerless Voltage-Tripler Power Supply

Power Supplies

stand higher voltages are expensive, so there is a practical limit to what can be done.

BRIDGE RECTIFIER POWER SUPPLY

Figure 7.7 illustrates a bridge rectifier circuit.

Distinguishing Features

Four diodes (not necessarily diagramed in a diamond pattern) connected across the secondary (or principle secondary) of a power transformer. This secondary does not have a centertap.

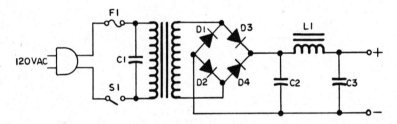

Figure 7.7   Bridge-Rectifier Power Supply

Uses

TV, radio, and other electronic equipment with comparable requirements.

Detailed Analysis

When the lower end of the secondary winding is negative, electrons flow via D2 to the load across the output terminals. From there the current returns through the filter L1-C2-C3 and D3 to the upper end of the transformer secondary. On the next half-cycle the polarities reverse, and the current flows from the upper end of the secondary winding through D1, and thence through the load and filter as before. However, this time it returns through D4 to the lower end of the winding.

VOLTAGE DIVIDERS

Power supplies which have to provide for more than one B voltage may do so by means of voltage dividers. A voltage divider consists of several resistors connected in series between the high and low sides of the output of a power supply. The current flowing through these resistors does the following:

(1) It stabilizes the power supply voltage.
(2) It discharges the capacitors when the equipment is turned off.
(3) It gives a voltage drop across each resistor which is proportionate to the resistance value.

In Figure 7.8 A, three resistors with values as shown are connected across a 240-volt supply. The total resistance is 40 kilohms, so the voltage divides proportionately as follows:

Resistor	Voltage Drop
10K	60V
15K	90V
15K	90V
40K	240V

In B a similar arrangement is shown connected across a 650-volt supply. However, the low side of the main circuit is not connected to the bottom of R5, but between R4 and R5, making this the zero reference. As a result the voltage at the bottom of R5 is *negative* (below zero) with respect to the zero point. This has the advantage that bypass and decoupling capacitors with lower working voltages can be used, which makes for greater reliability and less cost.

However, in transformerless power supplies there is a limitation on how much current can be supplied. If the voltage divider consumes too much it takes it away from the tubes for which it is supplied. In the examples shown, A draws 6 milliamperes, and B 10 milliamperes. In a TV set this may be more than can be tolerated. Consequently an arrangement similar to that shown at C may be used instead. This uses the voltage drop across a tube which draws considerable current (such as an audio-output tube) to provide two levels of voltage. The current passing through V4 is equal to the combined currents of V5, V6 and V7.

Two 100-volt tubes may be connected across a 200-volt supply in

Power Supplies

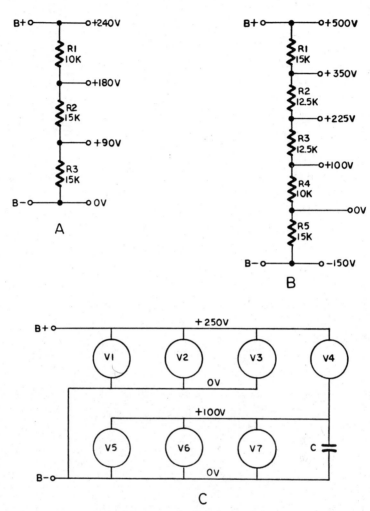

Figure 7.8 Voltage Dividers

series, provided they draw the same current. An example of this economical trick is shown in Figure 2.35.

VOLTAGE REGULATION

The voltage output of any of the power supplies discussed so far will vary according to the load and line voltage. In many cases this does not matter, but where a constant output is essential a *voltage regulator* circuit must be provided.

A voltage regulator consists of the elements illustrated in block form in Figure 7.9. It is connected to the output of a power supply (such as a bridge rectifier, Figure 7.7).

If the output voltage changes, the voltage at the movable arm of the "voltage adjust" resistor changes, and this change is sensed by the *comparator*, which monitors it constantly against the *reference*. The comparator passes on this information to the *error amplifier*, which in turn controls the *series regulator*.

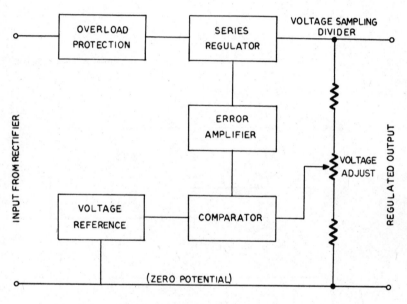

Figure 7.9 Voltage Regulation

In Figure 7.10 is shown a voltage which also contains an *overload protection* circuit (R1, Q1). If a short circuit occurs somewhere so that a very heavy current is drawn through R1, the base-emitter junction is forward-biased so that current can flow through Q1 and R5. The voltage drop across R5 is such that the base-emitter junction of Q2 (the error amplifier) is reverse-biased, so Q2 is cut off. As no current can now flow through R3, the forward bias on the base-emitter junction of Q3 disappears, and it also ceases to conduct.

When the overload is removed, so that normal current flows through R1, Q1 returns to its usual non-conducting condition, and Q2 and Q3

Power Supplies

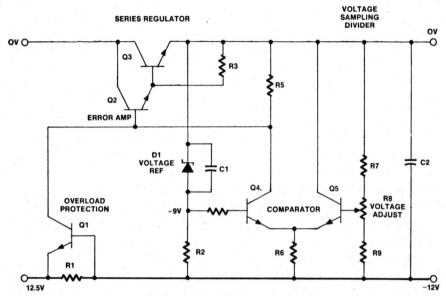

Figure 7.10 Voltage Regulator

become conductive again. R1 has a resistance value of .5 ohm or less, so normal current does not produce enough voltage drop to bias Q1 into conduction. C2, a large-value capacitor, absorbs noise and spikes so that operation is not interrrupted by transients.

D1 is a *zener diode* connected in series with a limiting resistor R2, and it gives a reference voltage of −9 volts at their junction. It also clamps the voltage on the base of Q4 at a value close to this. The base of Q5 is connected to the movable arm of R8. If the voltage at this point changes, due to a change in the output voltage of the power supply, the bias on Q5 changes. This changes the resistance of Q5, so that the current flowing through it changes.

Q4 and Q5 form a *comparator circuit*, and share R6. When the current through R6 changes (which it does when the current through Q5 changes) the voltage drop across R6 also changes. This changes the bias on Q4, so that its resistance changes in accordance with the increase or decrease of output voltage across R7, R8 and R9.

The change in voltage on the base of Q2 due to the change in voltage drop across R5 is amplified and applied to the base of Q3, causing this transistor to decrease or increase resistance to correct the rise or fall in the output voltage. Since the output of the power supply is dependent on the voltage at the movable arm of R8 the value of the output voltage can be adjusted by moving this control.

This is a very sensitive circuit, and is used extensively in well-regulated power supplies. The identical circuit will also be met where Q4 and Q5 are a dual-triode tube, Q2 is a pentode, Q3 a triode (or two or more triodes in parallel) and D1 is a gas glow-tube.

Circuit Variations

A circuit which is also often used is shown in Figure 7.11. This combines the comparator and error amplifier in one tube, V3. Changes in voltage at R9 change the voltage on V3's grid, thereby changing its resistance, and consequently the current through V3 and R5. The resultant change of plate voltage appears on V2's grid, so that when the output voltage rises V3's resistance increases to correct it. The neon glow-tube V1 clamps the cathode of V2 at approximately half the output voltage.

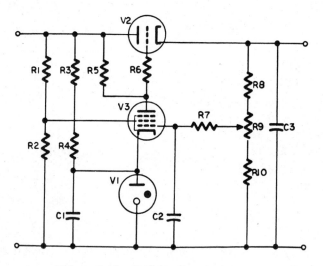

Figure 7.11 Voltage Regulator

Very much simpler and more economical circuits were illustrated in Figure 7.4, where Q1 and Q2 are also voltage regulators. Here only one transistor in each supply is doing the work of all four in Figure 7.10. Of course, the regulation is nowhere near as good, but it is much more economical, and adequate for hi-fi and the like.

In the case of Q1, D5 is also a zener diode, which clamps the base of Q1 to approximately 12 volts. Any variation in output voltage is felt by the emitter, changing the emitter-base bias, and controlling the

Power Supplies

conductivity of Q1. R6 is a variable resistor. By adjusting R6 the 9-volt output can be set to the exact value.

In the case of Q2 no zener-diode reference is provided, and consequently regulation is less efficient. Line-voltage variations are not compensated for, only load variations. The 40-volt supply is not regulated at all.

HIGH-VOLTAGE POWER SUPPLIES

Low-voltage power supplies, such as those we have been discussing, provide voltages up to around 500 volts. Applications needing higher voltage are served by high-voltage power supplies.

High-voltage power supplies fall into two classes: those where power is required, and those where voltage is the main consideration.

HIGH-VOLTAGE POWER SUPPLY FOR TRANSMITTER

Figure 7.12 illustrates one of the first type, to provide power for a transmitter. It is a full-wave power supply which differs only from a low-voltage supply in tube and transformer sizes. Its distinguishing features and mode of operation are the same as those for Figure 7.3, except for two additional features.

L1 is a *swinging choke*. This is a choke operated with the core close to saturation, so that the inductance varies with the current (see discussion of Magnetic Amplifiers in chapter 3). It has a smoothing effect on the DC pulses coming from the rectifier tubes.

R1 is a bleeder resistor. By providing a constant current drain it stabilizes the output voltage, and also discharges the filter capacitors when the power is switched off.

Note that there are *two* transformers. T1 is called the *plate transformer*. It is not uncommon for the AC potential across its secondary to be 2000 volts, or 1000 volts across each half.

T2 is the *filament transformer*. It provides the current to heat the filaments of V1 and V2. Since the DC potential on these filaments may be 1000 volts or more this transformer has to be well insulated. The tungsten filaments of these tubes will usually draw a heavier current than those used in smaller power supplies.

The tubes themselves are often heavy-duty mercury-vapor rectifiers, such as 816s.

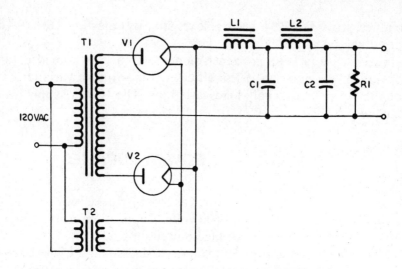

Figure 7.12   High-Voltage Power Supply for Transmitter

Capacitors C1 and C2 are generally of much smaller capacitance than those used in low-voltage power supplies. Electrolytics are not able to withstand the higher voltage, so these will probably be oil-bath types. Because they can only be around 5 or 10 microfarads at most, chokes with considerably higher inductance values have to be used.

HIGH-VOLTAGE POWER SUPPLY FOR CATHODE-RAY TUBE

Cathode-ray tubes (CRTs) are mostly used in oscilloscopes, although a TV picture tube is really a CRT also. However, picture tubes are classed by themselves in this chapter. CRTs require a high anode-to-cathode voltage. In many cases this may be as much as 12 kilovolts, but the current drawn is very small. Consequently the type of power supply shown in Figure 7.13 is used. TV HV supplies are similar, but are discussed in the next section.

Distinguishing Features

The output of an oscillator V1 is stepped up by a high-voltage transformer T1. The secondary voltage is rectified by diodes and filtered by resistors and capacitors, and the DC output is applied to the

Power Supplies

CRT, positive going to the anode, negative to the cathode and grid. A sample is often fed back to a comparator which controls the oscillator.

Uses

Mainly in high performance oscilloscopes or similar equipment.

Detailed Analysis

In Figure 7.13 V1 and the primary of T1 form a Hartley oscillator, in which the inductor resonates with the distributed capacitance of the circuit (see chapter 4). The supply voltage for this oscillator comes from the low-voltage supply of the oscilloscope. The frequency of oscillation will be in the order of 50 to 100 kilohertz. Filtering such a frequency is easier, and requires much smaller capacitors, which is an important consideration at high voltages.

Two voltages are supplied by this circuit. A positive voltage of 10 kilovolts is required for the CRT anode, and a negative voltage of 2 kilovolts for the cathode and grid.

The anode voltage is obtained by using a voltage tripler (see Figure 7.6). However, the higher frequency makes filtering easier, so smaller values are required for C5, C6, C7 and C9, and the choke is replaced by R10 and R11.

The cathode voltage, being much lower, is provided by the half-wave rectifier circuit consisting of D1, C4 and C8. Because of the higher frequency and the very small load involved this simple filtering is sufficient.

R7, R8 and R9 form a voltage divider in which R9 is very much larger than the other resistors. The voltage sensed at the movable contact of R8 is only a small percentage of the total. This voltage is fed to the right-hand grid of the dual triode V2.

The right-hand cathode of V2 is connected to the negative B supply (usually −150 volts). This supply is well regulated, so the cathode voltage is constant. Consequently any change in the grid voltage is followed by a change in the plate voltage. This *error voltage* is coupled directly to the left-hand grid, and the resultant amplified change of plate voltage appears on the screen grid of V1.

This in turn changes the amplitude of the Hartley-circuit oscillations. If the negative voltage on R8 decreases, the right-hand plate voltage decreases, the left-hand grid voltage decreases, the left-

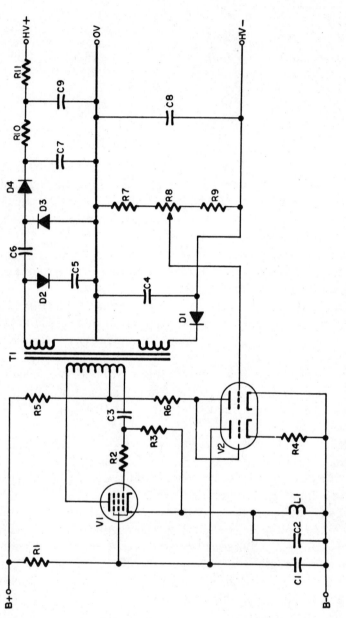

Figure 7.13 High-Voltage Power Supply for CRT

Power Supplies

hand plate voltage increases and the screen-grid voltage of V1 increases. This causes the oscillations to increase in amplitude, so that the high voltage rises.

The high voltage is set to the exact potential required by adjusting R8. L1 is a choke to keep the oscillator signal from coupling into the power supply (whence it would be distributed all over the set). R1, R5 and R6 are voltage-dropping resistors. R4 is the bias resistor for the left-hand half of V2; C1 and C2 are bypass capacitors.

Circuit Variations

In some models solid-state diodes may be replaced by high-voltage vacuum-tube rectifiers. V1 and V2 also may be transistors. The oscillatory circuit may be any suitable type instead of a Hartley. Part of the high voltage may go to the CRT intensity control. The anode or cathode may have independent supplies.

Finally, smaller or more economical equipment may obtain a sufficient high voltage from a secondary on the low-voltage power supply transformer.

TV HIGH-VOLTAGE SUPPLY

Figure 7.14 shows a typical high-voltage supply for the picture tube of a TV set.

Distinguishing Features

    a. One or more pentode or beam-power tubes, often with the plate connection to a cap on top of the tube, as symbolized in V1.
    b. A diode with indirectly heated cathode (V2), frequently having an adjustable inductor in its plate circuit.
    c. A transformer, usually an autotransformer as shown, with connections to V1, V2, the picture-tube horizontal yoke coils, a high-voltage rectifier (V3), and the picture tube itself (V4).
    d. In color TV sets an additional tube, usually a triode (V5), is added for regulation.
    e. Any of these tubes may be replaced by an equivalent semiconductor.

Uses

TV high-voltage section.

Detailed Analysis

Figure 7.14(A) illustrates a circuit which has become pretty well standardized for black-and-white sets. It has some similarities to the circuit in Figure 7.13. For example, the drive is from an oscillator and, because of the higher frequency, filtering is much simpler.

DC Subcircuit: Electron flow is from B– via R1 and V1 to T1, through T1 to V2, and via V2 and L1 to B+; also, from B– via R4 to V4, and via V4 and V3 to T1, and thence via V2 and L1 to B+.

AC Input Subcircuit: The horizontal oscillator (not shown) drives V1, which is a power amplifier, often called the horizontal-output tube, and similar to an audio-output stage. However, the oscillator frequency is 15.75 kilohertz, so it is a little too high for most people's hearing.

The oscillator signal is a modified sawtooth. V1 is biased so that current does not begin to flow until the sawtooth is halfway up its rise. At this point the electron beam in the picture tube is halfway along one of the horizontal lines. In other words, no current is flowing in the horizontal yoke coils to move it one way or the other. But when V1 starts conducting, an increasing current in the yoke coils causes a magnetic field to build up which drives the electron beam at a steady rate to the right-hand edge of the picture-tube screen.

At this point the peak of the sawtooth arrives, and the grid voltage on V1 dives down to its starting value. V1 cuts off abruptly, and the energy stored in the magnetic field of the yoke coils is suddenly released in a tremendous surge of current in the opposite direction.

The reverse current whips the beam back to the left-hand side of the screen (during the brief period when it is blanked out), and also cuts off V2 (the damper tube) by making its cathode more positive than its plate.

However, the positive surge now reverses in a negative direction, which starts the damper tube conducting again. This puts a heavy load on the resonant circuits, damping out the high-amplitude oscillations so that the energy is now released through the yoke coils in such a way as to allow the beam to return to the center of the screen. The circuit is designed so that the beam moves at a steady rate across the screen, with a smooth transition at the midpoint.

Power Supplies

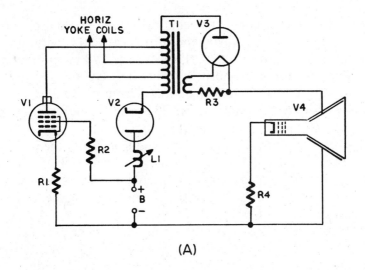

(A)

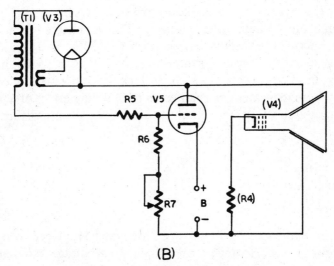

(B)

Figure 7.14 TV High-Voltage Supply

The high voltage is generated by the first surge of current, when V1 cuts off. Since V2 is also cut off, loading on T1 for the moment is minimal. The pulse amplitude can be as much as 2 or 3 kilovolts. T1 steps this up to 12 or 15 kilovolts for black-and-white sets, 25 kilovolts approximately for color.

These positive pulses are rectified by V3, and fed to the anode of V4.

The inner and outer conductive coatings on the envelope of the picture tube are connected as shown, and form a capacitor (with the glass as dielectric) that, with the internal resistance of V4 in series with R4, is sufficient to smooth the positive pulses into DC.

Filament voltage for V3 is obtained from an additional winding (usually only a turn or two) on T1. R3 is a low-value resistor to protect the filament, and is often omitted.

L1 is the *linearity control*. Adjusting it modifies the waveform of V1's plate current and also V2's discharge waveform. Another variable inductor (not shown) is usually connected in parallel with part of T1. This is the *width control*. It has the effect of varying the turns ratio of T1, and hence the step-up ratio.

In Figure 7.14(B) is shown an additional circuit used in color TV sets. V5 is a voltage regulator. Its grid voltage is set by adjusting R7. This varies the conductivity of V5, and so changes its loading effect on the high-voltage power supply.

The picture-tube load on the high-voltage power supply is varying all the time with changes in picture brightness. A bright scene increases the load on the high voltage supply. This is reflected in a fall in the voltage across the voltage divider R5, R6 and R7. As a result V5's grid voltage drops and the tube's internal resistance rises, so that it loads the high voltage less. When the scene is dark the opposite happens. In this way V5 compensates for changes in the loading effect of V4, so as to maintain a constant voltage on its anode. This is important for good color.

Circuit Variations

All the tubes except the picture tube may be replaced by equivalent semiconductors. However, the single high-voltage rectifier tube is usually replaced by a solid-state voltage doubler, tripler or even quadrupler. (A quadrupler is a tripler with an additional diode-capacitor pair, as mentioned in the discussion of Figure 7.6.) This gets the necessary high voltage from the flyback transformer, even though the input voltage to it—from a transistor—is much lower than would have been the case with a tube.

Regulation may also be performed on the primary side of T1. In a circuit somewhat similar to Figure 7.13, V1 is controlled to maintain a constant voltage output.

DC INVERTERS

So far we have been mostly concerned with converting AC power to DC power. Circuits for converting DC to AC are also required. Since transformers will not operate on DC it is necessary to convert DC to AC before it can be applied to the primary of a transformer to obtain a voltage step-up.

Oscillators convert DC to AC, of course, but are not generally suitable for power supplies unless the power requirements are small, as in high-voltage power supplies (Figure 7.13). Where the load is heavier some device must be used that "chops" a considerable DC current into pulses.

Such devices are *electronic* or *mechanical*, and are used in circuits called *DC inverters*. Electronic devices include *thyratrons, ignitrons* and *silicon-controlled rectifiers* (SCRs). Mechanical devices consist of *choppers, vibrators* or *dynamotors*.

In Figure 7.15 is shown a DC inverter circuit with a thyratron. The same circuit could also use an ignitron or an SCR (see under Circuit Variations).

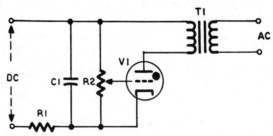

Figure 7.15   DC Inverter with Thyratron

Distinguishing Features

A gas-filled triode (note black dot, which indicates gas content instead of vacuum) is connected in a circuit with a DC input, and an AC output.

No DC operating voltages are provided from a power supply, as this *is* the power supply.

Uses

Heavy-duty power supplies converting DC to AC.

Detailed Analysis

In Figure 7.15, before any voltage is applied to the input, C1 is not charged. When power is turned on all the current flows into C1 at first, so no voltage appears across V1 and the primary of T1. Consequently V1 does not conduct. The voltage builds up across C1 at a rate determined by the R1—C1 time-constant. When it reaches the value that makes V1 fire, the gas in the thyratron ionizes, and V1 goes from being a very high resistance to a very low one. This enables C1 to discharge practically instantaneously through V1 and T1's primary. In fact, because the primary of T1 tries to maintain current flow even after C1 is totally discharged, there is momentary overshoot which serves to reverse the voltage across V1, so that it switches back to a non-conducting state. The cycle is now repeated as the input voltage again begins to recharge C1 through R1.

R2 is used to adjust the voltage on the grid of V1 for best performance. The chopped DC current passing through the primary of T1 induces an AC output in the secondary. The output voltage depends on the turns ratio of T1.

Circuit Variations

Instead of a thyratron, this circuit could be designed with an ignitron or a silicon-controlled rectifier (SCR). The symbols for these are given in Table I, chapter 1. The ignitron would be connected with its anode and cathode terminals like V1, and its igniter as if it were a grid. Ignitrons are capable of conducting currents of thousands of amperes, and have many applications in industrial operations. An SCR is a semiconductor equivalent of a thyratron, so that it is also called a thyristor. It is a four-layer diode with an external connection that acts as a gate to control its switching level, and therefore the amount of current it can pass. This makes it more flexible than the thyratron.

Dynamotors are used to convert DC to AC or to DC of a higher voltage. They are essentially DC motors with a secondary winding on the armature from which the output is obtained, and so are really rotating transformers.

Choppers and Vibrators are electromechanical devices for interrupting the flow of current, operating in the same way as a buzzer or telephone bell. At one time all automobile radios obtained their power from the car's battery by using a vibrator in a circuit similar to

Power Supplies

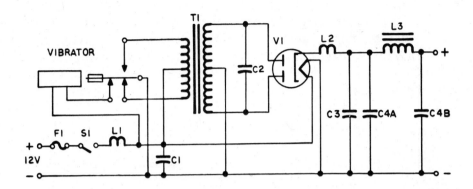

Figure 7.16   Vibrator Power Supply

Figure 7.16. Although modern car radios are solid-state it's worth noting this power supply because the principle of chopping DC and converting it to AC is well illustrated by it.

It is readily identified by the vibrator, which consists of an electromagnet, symbolized by the rectangle, and a reed. The reed rests against the two lower contacts (arrows). When S1 is closed current flows from the negative side of the 12V supply to the reed. Some of the current then flows through the lower half of the primary of T1 (via the right-hand contact) and back to the positive side of the supply from the centertap. The rest flows via the left-hand contact, through the electromagnet, and back to the positive side of the supply.

The electromagnet, being magnetized by the current, attracts the armature on the reed. This swings the reed over to the upper contact, breaking the connection with the two lower contacts. Current now flows in the upper half of T1's primary and back via the centertap, but in the opposite direction to that which flowed in the lower half. However, the current flowing in the electromagnet is now cut off, its magnetic field collapses, and the reed returns to its initial position, where the cycle starts again.

You can see that the alternate surges of current in opposite directions in T1's primary are an AC current, which is stepped up by the secondary, after which it is rectified and filtered by a full-wave power supply similar to that in Figure 7.3.

The cathode of V1 is indirectly heated by the 12-volt battery supply, as are all the other tube cathodes.

C2 is a *buffer capacitor*. It is used to reduce voltage surges ("spikes") that might damage the following parts. L2 and C3 together form a *hash suppressor* that filters out RF noise from the vibrator. L1 and C1 do the same for RF noise picked up by the car's wiring (such as sparkplug noise).

Thyratrons as Industrial Power Controls

Thyratrons are also used to control AC power for machinery by means of a phase-shifting circuit such as that in Figure 7.17.

The AC power source is connected to the anode and cathode of V1 as shown. The transformer T1 is also connected to the power source. Because of the centertap the polarity of the signal at the upper end of the secondary winding is opposite to that at the lower end.

The phase of the signal applied to V1's grid, relative to that of the signal on the anode, depends on the setting of R1, which forms a phase-shifting network with L1. If both signals are of the same phase, positive excursions will be applied simultaneously to anode and grid, and the thyratron will conduct for the maximum length of time until both signals swing negative, as illustrated in the left-hand set of waveforms.

However, if the phase of the grid signal is shifted, as shown in the right-hand set of waveforms, the period of conduction is shortened to the length of time during which the positive excursions overlap, so that the output power is reduced accordingly. R1 shifts the phase up to 90 degrees, with a corresponding variation in output from full power to zero.

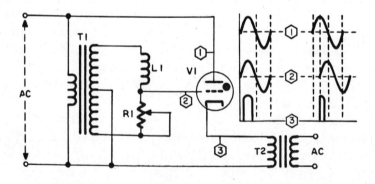

Figure 7.17   Thyratron Power Control Circuit

Diacs and Triacs

A *diac* is a two-terminal, transistor-like device which switches to a conducting state when the applied voltage reaches a certain level, typically three volts, regardless of polarity. Its major application is in conjunction with a triac to produce AC phase-control circuits, as explained below.

A *triac* is a modified PNPN device having five layers, that can switch on for either polarity of an applied voltage. It is the AC equivalent of an SCR, so it is employed as an AC light dimmer, motor-speed controller and so on.

Triac Power-Control Circuit

Figure 7.18 shows a very simple, inexpensive triac circuit used for dimming a light. At the beginning of each half-cycle of AC the diac and triac are not conducting. A small current (not enough to light the lamp) flows through R1 to charge C1. The voltage across C1 eventually reaches the level that causes the diac to conduct and discharge C1. This level is called the *breakover voltage* ($V_{(BO)}$) of the diac. The current now flowing through the diac switches the triac on, and a much heavier current flows through the lamp, so it lights. As the AC reverses polarity, the triac and diac switch off, and C1 begins to charge through R1 as before, but with opposite polarity. On reaching $V_{(BO)}$ the diac and triac switch on, and the lamp lights again.

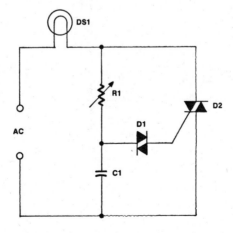

Figure 7.18 Basic Triac-Diac Light Dimmer Control Circuit

The time C1 takes to reach $V_{(BO)}$ depends on the setting of R1. With a higher resistance in the circuit it takes longer, and vice-versa. The length of time that D1 and D2 are conducting can, therefore, be adjusted so that full current flows in the lamp for a longer or shorter part of each half-cycle. This causes the lamp to be brighter or dimmer accordingly.

Similar circuits are used to control motor speeds, TV deflection systems, home appliances, automotive applications, and many more.

TROUBLESHOOTING TABLE FOR POWER SUPPLIES	\multicolumn{8}{c}{SYMPTOMS}
	No output	Low voltage	Hum or ripple	Video poor (TV)	Sound bars (TV)	No picture (TV)	No sound (TV or radio)	Vert. sync. critical (TV)
Line voltage low		X		X				
Line voltage zero	X					X	X	
Fuse blown, surge resistor open	X					X	X	
Power transformer open	X					X	X	
Power transformer - shorted turns			X					X
Rectifier(s) weak		X	X	X		X		X
Rectifier(s) open or shorted	X					X	X	
Filter capacitor(s) old, open, leaky		X	X	X		X		X
Filter capacitor(s) shorted	X					X	X	
Choke or series resistor open	X					X	X	
Choke shorted			X					X
Audio output tube defective*				X		X		
Audio decoupling capacitor defective*					X	X		
High-voltage power supply defective						X		

*Series-parallel system.

8
MODIFIERS

The signals used in electronic circuits usually have to be tailored to fit the purpose for which they are required. This tailoring is done with a variety of standard circuits which reduce the amplitude, change the shape, separate, or otherwise modify a signal.

Attenuators and Pads

The distinguishing feature of attenuators and pads is that they consist of resistors. Attenuators are resistive networks that reduce the strength of a signal without introducing appreciable distortion. A pad is a resistive network that attenuates a signal without introducing an impedance mismatch. It is also used to match dissimilar impedances to avoid unnecessary signal attenuation, in which case it is called a minimum-loss pad.

Fixed pads are either *symmetrical* or *asymmetrical*. In the first the input and output impedances are the same, in the second they are not. They are also *balanced* or *unbalanced*, depending upon whether the two sides are alike or unlike with regard to a common reference (usually ground). This means that an unbalanced pad is one with one side grounded or at zero potential, as in most circuits, while a balanced pad has neither side grounded.

In Figure 8.1 the first unbalanced pad is called an *L-pad* because the arrangement of its resistors resembles an upside-down L. Its balanced counterpart is a *U-pad*, because it looks like a U lying on its side. The resistors drawn horizontally are *series* resistors; the vertical are *shunt* resistors. These are impedance-matching pads, and are also called *minimum-loss* because they do not cause much attenuation.

The next unbalanced pad is a *T-pad*, and its balanced version is an *H-pad*. These pads are used for attenuation. They should be connected between similar impedances.

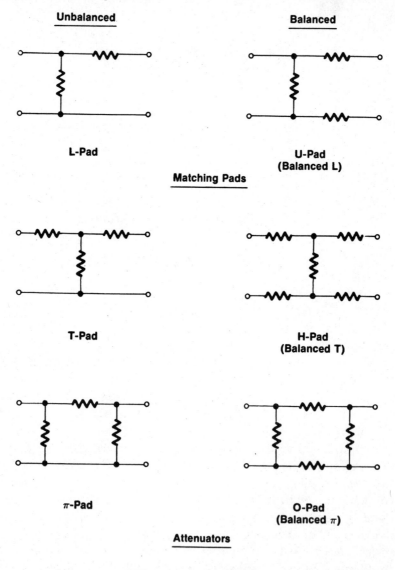

Figure 8.1 Fixed Pads

The bottom two pads are also for attenuation, not for impedance-matching. The π-*pad*, like the π-*filter*, is named for its resemblance to the Greek letter π (pi).

Figure 8.2 shows T and H pads with resistors shunted across the series resistors. They are called *bridged-T* and *bridged-H* pads.

Variable pads are also illustrated in Figure 8.2. These are designed so that the attenuation can be varied while maintaining the impedance match. In a simple potentiometer the amount of resistance shunting the device connected to it varies with the setting. In the examples shown, one resistor increases as the other decreases to maintain a constant impedance.

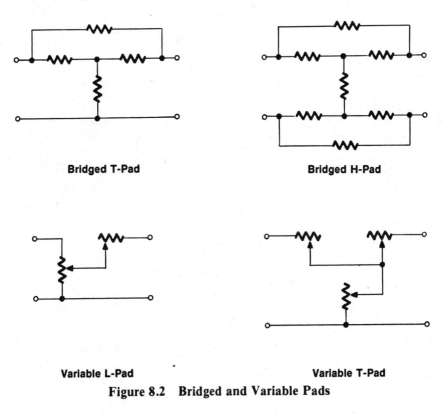

Bridged T-Pad   **Bridged H-Pad**

Variable L-Pad   **Variable T-Pad**

Figure 8.2   Bridged and Variable Pads

Filters

A filter is a selective network designed to pass signals within a certain frequency range ("pass band"), while attenuating those of other frequencies. Filters are categorized as *low-pass, high-pass, bandpass* and *bandstop*. The distinguishing feature of all of them is the use of inductor-capacitor (L—C) circuits. A low-pass filter passes lower frequencies but attenuates higher ones. Figure 8.3 shows six low-pass filters. They look somewhat like the pads in Figure 8.1, except that all are unbalanced. In each there is a *series inductance* and a *shunt capacitance*. The inductance has more reactance to higher frequencies, while the capacitance has more reactance to lower ones. Consequently,

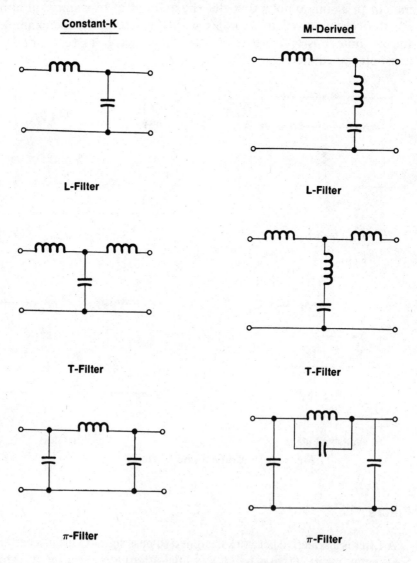

Figure 8.3 Low-Pass Filters

higher frequencies find an easier path through the capacitance, and so are bypassed, but the inductance offers less reactance to the lower frequencies which therefore are not bypassed.

Constant-k filters are simpler than *m-derived filters*, and do not have as sharp a cutoff. For many purposes this is no disadvantage. The power supplies discussed in chapter 7 make extensive use of low-pass filters, especially filters to eliminate 60- and 120-hertz AC, while

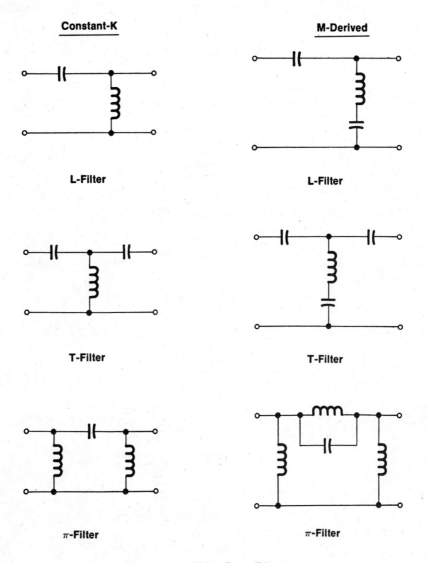

Figure 8.4 High-Pass Filters

passing DC (which in this connection can be regarded as ultra-low-frequency AC).

In the m-derived filters the addition of shunt inductances in the first two makes, with the shunt capacitance, series-resonant circuits having a very low shunt impedance to frequencies just below the cutoff frequency. This results in a steep drop in the frequency response below the cutoff frequency. The circuit in the third filter is a parallel-resonant circuit, and so offers a high series reactance to frequencies below cutoff.

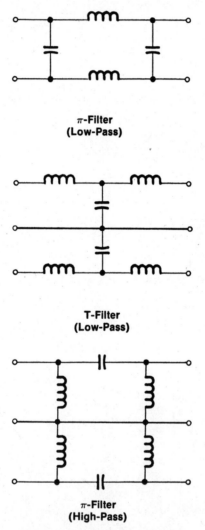

π-Filter
(Low-Pass)

T-Filter
(Low-Pass)

π-Filter
(High-Pass)

Figure 8.5   Balanced Filters

A *high-pass filter* passes higher frequencies but attenuates lower ones. Figure 8.4 shows six high-pass filters. They are the counterparts of those in Figure 8.3. In each one there is a *series capacitance* and a *shunt inductance*. The lower frequencies find an easier path through the inductance, and so are bypassed, while the capacitance offers much less reactance to the higher frequencies, which are not bypassed.

The distinctions between k- and m-derived filters are also similar to those that apply to low-pass filters.

The filters in Figures 8.3 and 8.4 were unbalanced filters. Examples of balanced versions are shown in Figure 8.5.

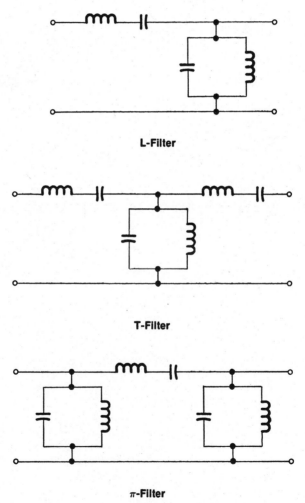

L-Filter

T-Filter

π-Filter

Figure 8.6   Bandpass Filters

Bandpass filters allow signals with the desired frequency to pass while rejecting all others. The width of the band depends on the selectivity of the circuit, which can be made broader by increasing the resistance in it. Figure 8.6 illustrates three typical bandpass filters of the constant-k type. They resemble the unbalanced pads shown in Figure 8.1, with series-resonant circuits substituted for the series resistors and parallel-resonant circuits replacing the shunt resistors. Series-resonant circuits offer a very low reactance to signals at and near the resonant frequency, but parallel-resonant circuits are the

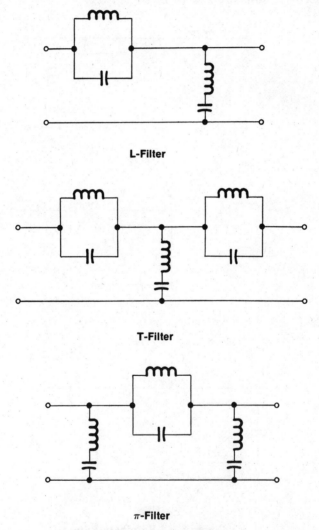

L-Filter

T-Filter

π-Filter

Figure 8.7 Bandstop Filters

Modifiers

opposite. To signals of other frequencies the series-resonant circuit offers a much higher reactance, the parallel-resonant one much lower. Consequently signals within the pass band find an easy path along the upper part of the circuit, and are not bypassed, while other frequencies are severely attenuated.

Bandstop filters are illustrated in Figure 8.7. If you compare this with Figure 8.6 you will see that they are counterparts. The series and parallel-resonant circuits have changed places. As a result, bandstop filters discriminate against the selected frequency instead of favoring it.

Circuit Variations

Since inductors are not readily adaptable to integrated-circuit techniques, which have become extremely important in recent years, a new class of *active filters* has been developed, constructed with resistors, capcitors and integrated-circuit operational amplifiers (see chapter 2). Figure 8.8 shows the four basic categories.

Integration and Differentiation

The two circuits in Figure 8.9 each consist of a capacitor and resistor only, which is what distinguishes them from the attenuators and filters we've just considered.

An *integrator* consists of a capacitor charging through a resistor. Its output represents the time integral of its input or, in other words, the output voltage is proportional to the time constant of the R-C combination. When a pulse with a waveform similar to the upper outline is applied to the input it is changed so that the output waveform is like the lower outline. This happens because the capacitor cannot charge instantly to the voltage of the pulse, but must build up the voltage at a rate determined by the flow of current through the resistor.

This circuit is used in television receivers to build up a voltage to trigger the vertical oscillator by accumulating the vertical synchronizing pulses that occur between each field (frame). These pulses have a long duration compared to the time interval between them, so that each pulse adds its voltage to that of its predecessor before the capacitor has had time to discharge, as shown in the dashed outlines in Figure 8.9.

Integrators are used in a similar way in counters, frequency dividers and the like. They may be compared to low-pass filters.

A *differentiator*, on the other hand, resembles a high-pass filter. This circuit consists of a capacitor and resistor arranged as shown. When a voltage is applied to the input, current flows to charge both sides of the

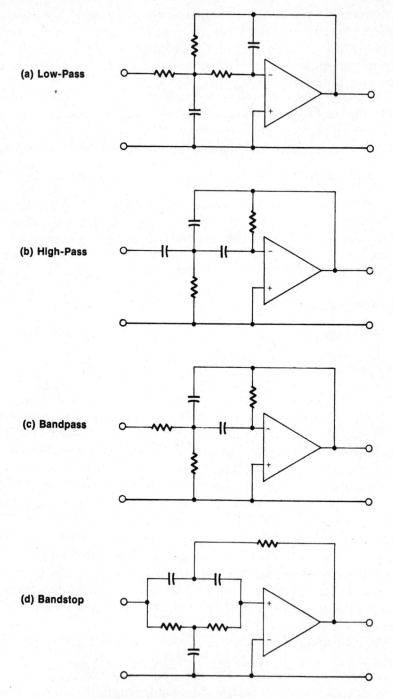

Figure 8.8 Active Filters

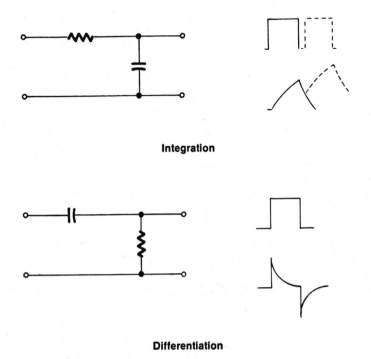

Figure 8.9 Integration and Differentiation

capacitor. As the capacitor has no charge at first, the initial current is high, tapering off as the capacitor becomes charged. This current flowing through the resistor produces a corresponding voltage drop that has the waveform shown, with a sharp positive-going spike initially when the leading edge of the pulse arrives, and a negative-going spike, caused by the discharge current, at the pulse trailing edge.

In television receivers this circuit serves to change the vertical synchronizing pulses into sharp triggering pulses to maintain horizontal synchronization during the period between fields when these extra-wide pulses are being received.

Clippers (Figure 8.10) may be either series or shunt. The distinguishing feature is a diode. Those shown in the figure are semiconductors, but vacuum-tube diodes could be used in the same way. These diodes have been *biased* by placing a DC source where it applies a reverse-bias to the diode, such that this voltage has to be overcome before the diode can conduct.

In the *series-clipper*, current can flow through the diode whenever the anode voltage is positive with respect to the cathode (the right-hand

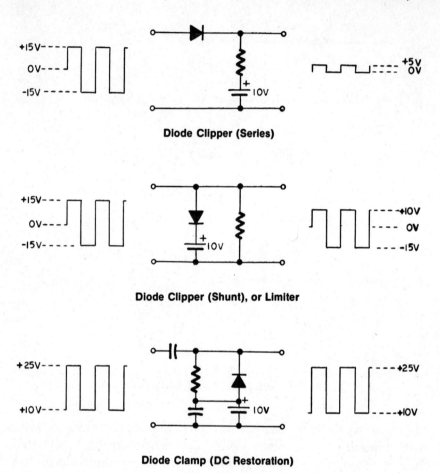

Figure 8.10 Clippers, Limiters and Clamps

side of the diode symbol, with the line). The DC source biases the cathode with a positive 10 volts. Consequently the diode can only conduct when the voltage on its anode exceeds 10 volts. If a square wave with positive and negative excursions (peaks) of 15 volts is applied to the input, the diode will conduct only during that part of the positive excursion that exceeds 10 volts, which is 5 volts. This gives an output of 5-volt pulses as shown. The rest of the input is blocked by the diode, and is said to have been clipped.

The *shunt clipper* has also been biased with 10 volts. A negative excursion of the square wave is not bypassed, as the diode does not conduct; consequently the whole negative 15 volts appears at the output. The diode does not conduct until more than 10 volts of the positive excursion appears on the anode of the diode. Voltage above this is bypassed, therefore only 10 volts of the positive excursion appears in the output.

The shunt clipper when used with a pulse train is also a limiter, since by clipping all the pulses to the same height, variations in amplitude (transients, overshoots, noise, etc.) can be removed. The series clipper does not do this, as in its case the bottom of the pulse is removed instead of the top.

The action of a clipper is the same for sine waves as for square waves. In fact, a sort of square wave can be produced by cutting off the tops of sine waves.

Reversing the polarity of diode and bias reverses the polarity of the clipping.

Clamp circuits are used for *DC restoration*. In the circuit shown in Figure 8.10 a train of pulses with voltage alternating between +25 volts and +10 volts is coupled through a series capacitor. This signal is equivalent to a train of 15-volt pulses riding on a 10-volt DC base. The capacitor cannot pass the DC portion, so only the train of pulses appears across the resistor.

The missing DC is restored by the DC source, which replaces the 10 volts that could not pass through the capacitor. This voltage is present on both sides of the diode. When a pulse arrives the voltage on the cathode increases to +25 volts; the diode does not conduct as its anode is still at 10 volts. However, any negative overshoot which takes the cathode below +10 volts causes the diode to conduct, and the overshoot is bypassed. In this way the pulse train is clamped to a 10-volt base.

If the DC source is omitted the pulse train will be clamped to a zero base. Other DC voltages may also be used. The shunt capacitor is provided to bypass the DC source so that no signal voltage is dropped across it.

9
LOGIC CIRCUITS

Logic circuits are circuits that answer questions with yes or no—not maybe! For instance, a circuit can be designed to answer the question "Can you start the car?" Suppose this circuit has been built inside a small box, as in Figure 9.1 (a). On the outside of the box are two switches. Above the left-hand one is the query "Do you have the ignition key?" If you do, you place the switch in the YES position. Above the right-hand switch the question is "Do you have gas in the tank?" If you do, you place this switch in the YES position also. Under the question "Can you start the car?" the lamp lights to indicate the answer is YES. If either switch had been in the NO position the lamp would not have lit. You can see why this is so from the circuit diagram at (b). Both switches must be closed (YES) before the current from the battery can flow through the lamp. This circuit is therefore called an AND circuit, because S1 *and* S2 must be on before DS1 can light.

Now take a different question, "Can you get a loan from the bank?" This circuit is also in a box like the first, as shown in Figure 9.2 (a). Above the left-hand switch the label asks "Do you have good credit?" When you place the switch in the YES position the lamp lights to tell you the bank says YES. It doesn't matter which position the other switch is in. However, if you place the left-hand switch in the NO position, the lamp will not light unless you place the right-hand switch in the YES position. Again, the reason is clear from the circuit diagram at (b). Closing either switch allows current to flow through the lamp.

Logic Circuits

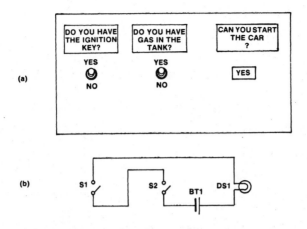

Figure 9.1 Principle of an AND Gate
(a) Input/Output Panel
(b) Schematic

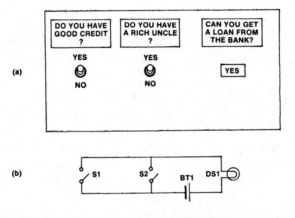

Figure 9.2 Principle of an OR Gate
(a) Input/Output Panel
(b) Schematic

This circuit is therefore called an OR circuit, because either S1 *or* S2 must be on before DS1 lights. (Of course, it doesn't hurt if both are on; the bank will like you even more!)

The AND and OR circuits are called *gates*. As we shall see in this chapter, there are two others called NAND and NOR. In each case a gate is a circuit having two or more inputs and one output. As we have just seen, the output (whether the lamp lights) depends upon the combination of the inputs (the positions of the switches). Other logic circuits are *inverters* and *flip-flops*. Together with amplifiers, oscillators, and some other circuits, they are the building blocks of digital computers.

By combining large numbers of these elementary units we can build computers to process all kinds of data, as long as we can make the switching automatic. This is done by using semiconductor switches (diodes and transistors). These are operated in one of two states: conducting and non-conducting. A transistor, for instance, will be saturated (fully conducting) or cut off, according to whether the voltage applied to its base is "high" or "low." The exact voltage is less important than that the two levels be sufficiently discrete to ensure reliable switching from one state to the other.

Inverter

Figure 9.3 illustrates an inverter. As you can see, it is a simple common-emitter amplifier. If the supply voltage is positive (with a PNP transistor, as shown) it requires a positive voltage on the base to make it conduct.

Without a positive voltage on the base the transistor-switch is off, or non-conducting. The voltage at A is said to be low or false, symbolized by 0. The voltage at B is, of course, the same as the supply voltage, so it is called high or true, symbolized by 1. When a positive voltage is applied at A the transistor switches to its fully-conducting state, so the voltage at B drops to zero. A is now high (1) and B is low (0).

The action of the inverter can be given in a table, as in Figure 9.3. This shows that when A is 0, B is 1, and vice-versa. The output is always opposite to the input. This type of table is a *truth table*. It is a graphic way of showing what a logic circuit does in response to various inputs.

The triangular symbol is, of course, the symbol for an amplifier. However, to make sure everyone understands it has an inverted output, the small circle is added.

A less common name for this circuit is a NOT gate (because the output is *not* the same as the input).

AND Gate

Figure 9.4 shows a semiconductor AND gate. The transistors Q1 and Q2 are connected in series, so that both have to turn on before

Logic Circuits

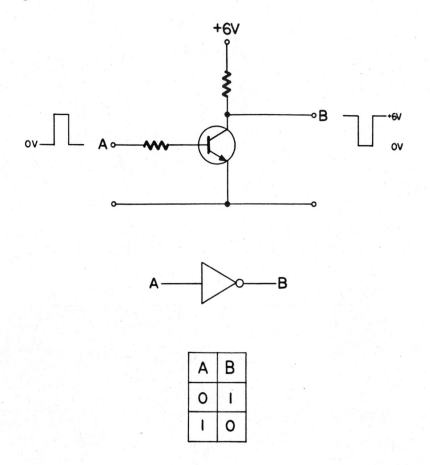

Figure 9.3 Inverter

current can flow through them. The collector of Q1 remains at the supply voltage until true signals are received at both A and B, when it drops to approximately zero. This turns Q3 off, and the voltage at C rises to the value of the supply voltage (Q3 is an inverter). The effect of different inputs is shown in the truth table. The symbol for an AND gate is given also.

OR Gate

The OR gate in Figure 9.5 is activated by a true input at either A or B, since the transistors are in parallel. When both transistors are off, the voltage at C will be zero. When either or both turn on, current

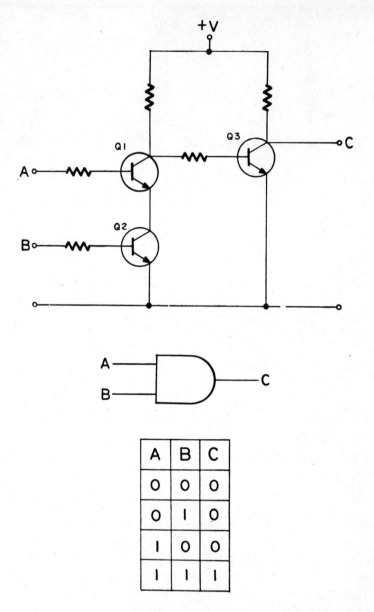

Figure 9.4 AND Gate

flowing through R2 causes the voltage at C to rise approximately to the value of the supply voltage. Consequently a true input at either A or B gives a true output at C, as shown in the truth table. Note the symbol for an OR gate.

Logic Circuits

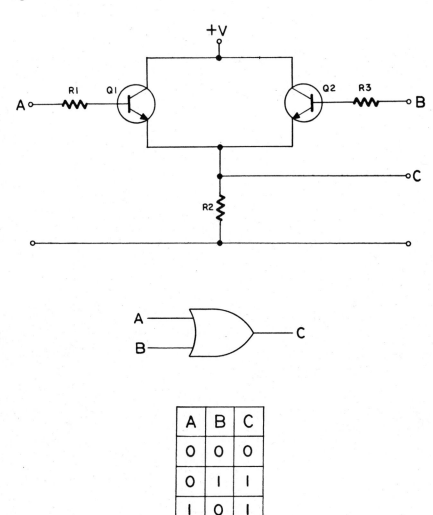

Figure 9.5 OR Gate

Compare the operation of these semiconductor AND and OR gates with the manually operated ones described above.

RTL NOR Gate

A NOR gate is an OR gate with a reversed output. In Figure 9.6 the

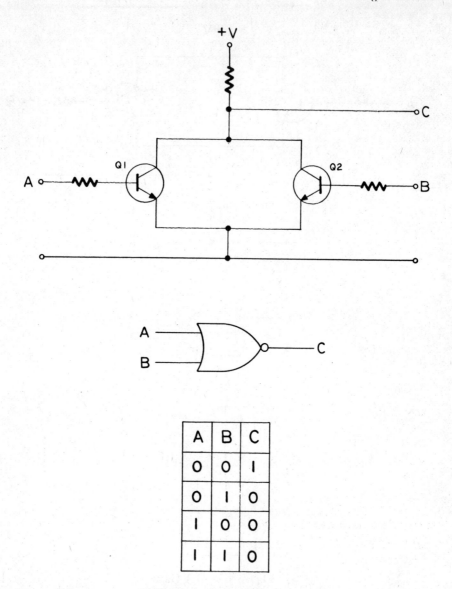

Figure 9.6 RTL NOR Gate

OR symbol now has a small circle added, which indicates the inversion. This could be done by adding the inverter of Figure 9.3. However, in this case, by taking the output from the *collectors* of Q1 and Q2 an inverted output is also obtained, as shown in the truth table.

This is one of the most widely used gates. RTL stands for resistor-transistor logic.

Logic Circuits

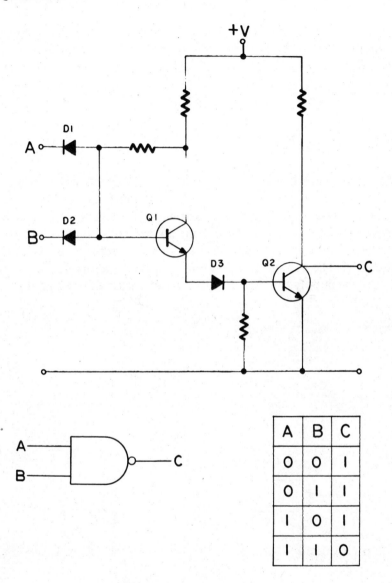

Figure 9.7 DTL NAND Gate

DTL NAND Gate

A NAND gate (Figure 9.7) is an AND gate with a reversed output. Note that the AND symbol now has a small circle added, which indicates the inversion. DTL stands for diode-transistor logic.

As long as either diode can conduct, Q1's base voltage is low, and Q1 is cut off. As Q1's emitter is at zero potential (with no current flowing through Q1) Q2 is also cut off. When true signals appear at A *and* B both diodes are reverse-biased and cannot conduct. Consequently Q1's base voltage rises, turning Q1 on. The voltage drop across Q1's emitter resistor turns Q2 on, resulting in a false output at C.

TTL NAND Gate

The TTL (transistor-transistor logic) NAND gate (Figure 9.8) is also widely used. It has a higher speed of operation than the DTL.

Q1 is a special transistor with two or more emitters (multiple-emitter transistor). When either or both emitters have a false input Q1 is cut off. The output from Q1's collector is essentially zero. As a result Q2 is also cut off. The voltage on Q2's emitter is zero, that on its collector the same as the supply voltage. In other words, with a false signal on its

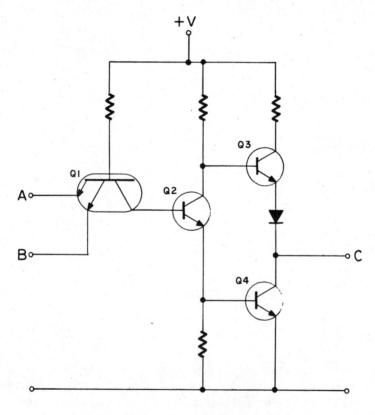

Figure 9.8 TTL NAND Gate

Logic Circuits

base Q2 has two outputs: false from the emitter, true from the collector. Q2 is therefore a *phase-splitter*.

Q3 and Q4 are connected in series (a "totem pole"). When Q2's collector is true the base of Q3 is true also, so Q3 is turned on. At the same time Q4 is turned off by the false signal from Q2's emitter. As a result the collector of Q4 is at the supply voltage (all the voltage is dropped across Q4), and the output at C is true.

When true signals are applied simultaneously to A and B Q1 turns on. This places a positive (true) voltage on the base of Q2, so its two output signals are now the opposite of what they were before. Consequently Q3 turns off and Q4 turns on. Now a false (zero) output appears at C.

The truth table and symbol for this circuit are the same as in Figure 9.7.

AND and OR Gates from NAND and NOR

Since the NAND and NOR gates just described are widely used today, AND and OR gates are often derived from them by adding

AND

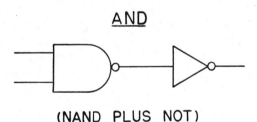

(NAND PLUS NOT)

Figure 9.9 AND Gate from NAND Gate

OR

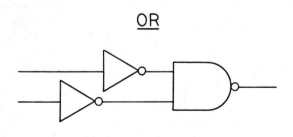

(NOT PLUS NAND)

Figure 9.10 OR Gate from NAND Gate

OR

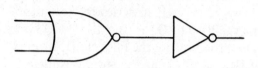

(NOR PLUS NOT)

Figure 9.11   OR Gate from NOR Gate

AND

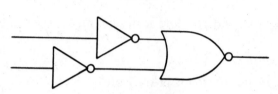

(NOT PLUS NOR)

Figure 9.12   AND Gate from NOR Gate

inverters to the output or input, as shown in Figure 9.9. In this way a NAND gate can be converted to an AND or an OR gate; similarly, with a NOR gate. In this way the same type of gate can be used throughout an entire digital system. The most popular type is the NAND gate. See also Figures 9.10, 9.11 and 9.12.

Flip-Flop

The foregoing circuits are somewhat like push-button switches: when you remove your finger from the button the bell stops ringing. This is all right for doorbells, but would be highly inconvenient for room lights, and also for many logic operations. To maintain the output in a given logical state after the input signal has been removed requires a *flip-flop*. The circuit acts like a light switch: when you turn it on it stays on until you turn it off.

Figure 9.13 shows a basic flip-flop circuit. It consists of two transistor inverters with their bases and collectors cross-connected. Let's assume that the input signal at S is low (0) initially. Q1 has a low voltage on its base, so it cannot conduct, therefore its collector is high (1). This high potential must also be present on Q2's base, so Q2

Logic Circuits

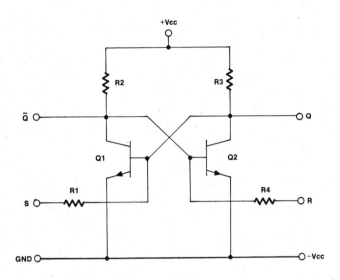

Figure 9.13 Basic Flip-Flop Circuit

presents a minimal resistance to ground, resulting in a low (0) state at the output, which is taken from its collector.

Now we apply a high (1) input signal at S, and this appears also on Q1's base. Q1 switches to its conducting state, and its collector potential falls almost to zero, or low (0). This low potential appears also on Q2's base, so Q2 ceases conducting (switches off), and its collector voltage rises to that of the supply voltage, so that the output is now high (1). This output potential also appears on the base of Q1, so Q1 remains in its conducting state even after the input signal at S is removed.

To reset the flip-flop to its original state we apply a positive (high) signal to the input R. This appears on the base of Q2 and returns Q2 to its conducting state, so its collector and the output go low (0). This low potential also appears on Q1's base, so Q1 reverts to its initial non-conducting state.

This type of flip-flop is called a Set-Reset, or S-R flip-flop, because the two inputs set and reset it. However, in the operation of a system including a flip-flop, it is usual to set and reset the circuit at specific times at which other circuits are also manipulated. This is done by adding clock pulses to the inputs so that the circuit only operates if both input and clock signals are present together. This can easily be done by putting AND gates in the set and reset inputs, so that no signal will appear on the bases of Q1 and Q2 unless both inputs of the AND gates are high at the same time. The circuit is then called a *clocked S-R flip-flop*.

Another type of clocked flip-flop which is widely used is the *J-K flip-flop*. As its logic diagram shows, it also has AND gates in its J and K inputs, so that a clock pulse must be received as well before the input signal can be passed on to the flip-flop. In Figure 9.14 the flip-flop is shown to consist of two cross-connected NOR gates, but its action is essentially the same as in the previous example.

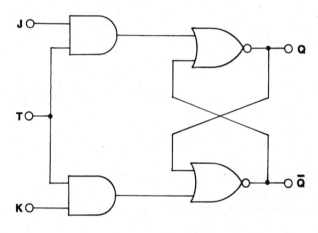

Figure 9.14   J-K Flip Flop

A typical standard circuit that is manufactured as an IC is a *J-K master-slave flip-flop*. As you can see in Figure 9.15, it is two of the circuits of Figure 9.14 "cascaded," as it were, so that the first drives the second, the output being taken from the slave stage.

At the application of a clock pulse when there is a 1 on the J input, the output Q will go to a 1. When the 1 is on the K input it resets Q to 0. If 1's appear simultaneously on both inputs, Q changes state regardless of what was on it before. The second available output, $\overline{Q}$, is always the opposite of Q.

Flip-flop circuits form the basic logic elements for storing information. In a computer, information is coded in *bits*. A bit is either a 1 or a 0, so you can see how a flip-flop can store these by being set in one state or the other.

Sets of flip-flops for storing data are called *registers*. The most common type of register is called a *shift register*. This consists of a number of flip-flops connected in series, the output terminals of one

Logic Circuits 247

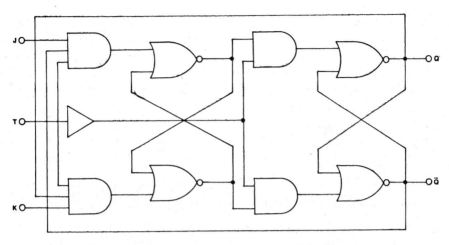

Figure 9.15 J-K Master-Slave Flip-Flop

being connected to the input terminals of the next. The same clock pulses are fed simultaneously to all the flip-flops in the register. When a clock pulse appears, information at the input of the first flip-flop is stored in that flip-flop. When the next clock pulse appears, this information is transferred to the second flip-flop, and a new input enters the first one, and so on down the line with each clock pulse. In this way several bits of information can be stored, one bit per flip-flop. This information can then be transferred out of the register, bit by bit, on command, from the output of the final flip-flop. Shift registers commonly store eight bits at a time, and are called *8-bit registers*. Other registers with different storage abilities may be met with, however.

Another version of the flip-flop circuit is described in chapter 12.

10
BRIDGE CIRCUITS

Bridge circuits are mostly used in measuring instruments, but they are also employed in sensing and controlling. They are the electronic equivalent of a pair of scales, in which some unknown quality is balanced against the known weights. When a bridge is "balanced" its two sides are at the same potential, so no current flows through the indicating device connected between them. Consequently it shows zero, or a "null." When a resistor, capacitor or inductor of unknown value is to be measured on a bridge, known values of resistance, capacitance or inductance are used as "weights" to balance the bridge, and from these values the value of the unknown is found.

Wheatstone Bridge

Bridges are frequently, though not always, shown in the diamond pattern of Figure 10.1. This is the Wheatstone bridge. It is the only bridge that can operate on DC, and then only for measuring resistance.

In Figure 10.1(a) resistors R1, R2 and R3 are bridge resistors, and R_x is the "unknown" resistor whose resistance is to be measured. In (b) the same layout is illustrated, but rotated 90 degrees clockwise. Either way is possible for this layout. R1 and R3 are a voltage divider, and so are R2 and R_x. Current from the battery flows through each divider, and the voltages at X and Y will be according to the ratio of the resistances.

Bridge Circuits

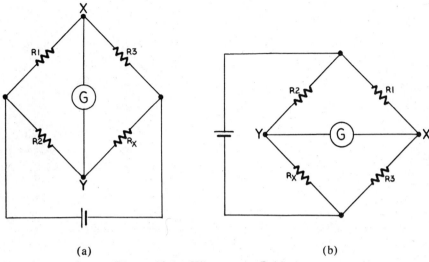

(a) (b)

Figure 10.1 Wheatstone Bridge

If R1 and R3 are equal the voltage at X will be half the battery voltage, and if R2 and R_x are equal the voltage at Y will also be half the battery voltage. X and Y will be at the same potential, so no current will flow between them. The galvanometer, G, connected between them will give no indication. We say the bridge is "balanced" or the galvanometer is "nulled," or indicates a null.

Since R_x is unknown R2 is made variable so that it can be adjusted to balance R_x. It may be an assemblage of fixed resistors, a potentiometer, or both. R1 and R2 together must provide resistance such that the right current flows through R2 and R_x to allow the voltage at Y to change readily as R2 is adjusted, so that an accurate reading can be made. The value of R2 is read from the dials of the selector controls, and is of course the same as that of R_x.

LC Bridge

The LC bridge is a Wheatstone bridge adapted for measuring inductances and capacitances. Since we are now measuring reactance we have to use AC instead of DC.

In Figure 10.2(a) the unknown inductance is balanced against a known inductance just as in the resistance bridge. However, the dials giving the reading are calibrated in henries, millihenries and microhenries instead of ohms.

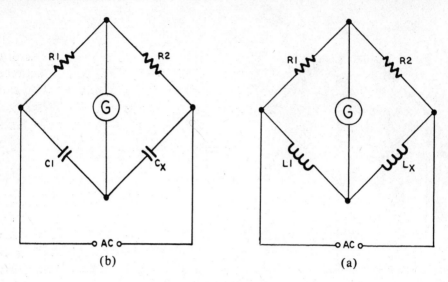

Figure 10.2   LC Bridge

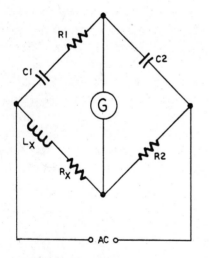

Figure 10.3   Owen Bridge

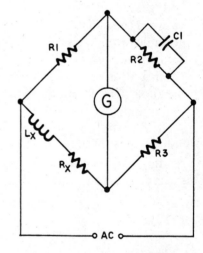

Figure 10.4   Maxwell Bridge

In Figure 10.2(b) a bridge for measuring capacitance is shown, operating in a similar way.

The function of R1 and R3 (or R2) is the same for all three bridges.

Owen Bridge

In the Owen bridge (Figure 10.3) R1 and R3 of the Wheatstone bridge are replaced by C1 and C2. This bridge measures inductance more accurately because it measures both the inductive (L_x) and resistive (R_x) components of an inductor. C1 and R2 are each variable, and must be adjusted to obtain a null. In another version a variable resistor in series with L_x is used instead of C1.

Maxwell Bridge

The Maxwell bridge (Figure 10.4) is very similar to the Owen, and is used for the same purpose. It is a partial return to the Wheatstone bridge, since R1, R2 and R3 have been modified only by the addition of C1 in parallel with R2. The resistive part of the inductance is therefore measured as in a Wheatstone bridge, but C1 adapts it for measurement of inductance also.

Hay Bridge

The Hay bridge (Figure 10.5) is similar to the Maxwell, except that C1 is now used in series with R3. This arrangement works better for inductors with large inductances.

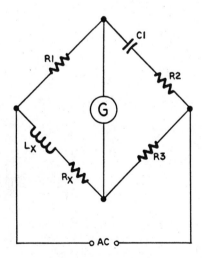

Figure 10.5   Hay Bridge

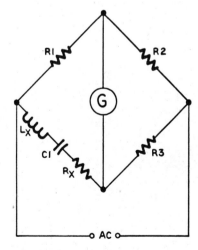

Figure 10.6   Resonance Bridge

Resonance Bridge

In this bridge (Figure 10.6) the capacitor C1 is tuned to resonance with the unknown inductor. This serves to measure its inductance, and at the same time leaves only the resistive element (at resonance a series-resonant circuit has no reactance), which is balanced by R3, as in the Wheatstone bridge.

Schering Bridge

The Schering bridge (Figure 10.7) measures capacitance. Its operation is similar to the Maxwell bridge, except that C2 replaces R2. The unknown capacitor has a resistive component in its reactance in the same way an inductor does.

Wien Bridge

The Wien bridge (Figure 10.8) is used for measuring frequency. It is similar to the Schering bridge, and can be used to measure capacitance. However, if instead of an unknown value of capacitance at C_x a known value is chosen to balance that of C1 at the frequency being measured, and if R3 is also adjusted to balance the resistive element of C_x, then the frequency of the AC signal applied to the bridge may be determined.

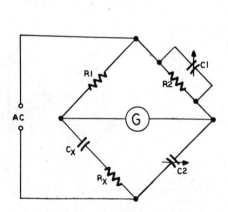

Figure 10.7 Schering Bridge

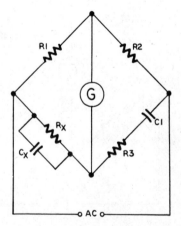

Figure 10.8 Wien Bridge

11
SYNCHRO SYSTEMS

Just as a thermostat senses a drop in temperature and signals the furnace to turn on, so a synchro generator senses a change in the mechanical position of something, and sends a signal which actuates the corrective action. Synchros are used extensively in industrial controls, in the guidance systems of ships, planes and rockets, and in other applications too numerous to list.

Servomechanisms

A household thermostat has a fixed output, and is called a *regulator*. A more sophisticated control system that adjusts its output automatically in accordance with its input is called a *servomechanism*. A well-known example is the governor on a steam engine, in which an increase in speed above that set results in a reduction in throttle opening, and vice-versa, to maintain an even rate.

Synchro Systems

A synchro system is a feedback system in which synchros are used. A synchro (sometimes called a selsyn) resembles a small electric motor. Three field coils mounted in a cylindrical case surround an armature (Figure 11.1(a)). The field coils are called *stator windings,* those on the armature are called *rotor windings*. However, when an alternating current of 60 or 400 hertz (depending on synchro type) is applied to the rotor winding the armature does not rotate. Instead its magnetic field induces voltages in the three stator windings. Because each stator

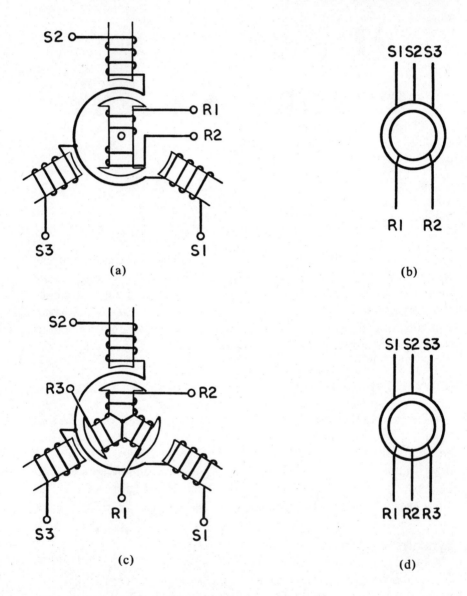

Figure 11.1 Synchro Construction

Synchro Systems

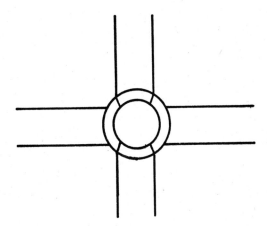

Figure 11.1 Synchro Construction (continued)

winding makes a different angle with the rotor winding a different voltage is induced in each.

These voltages rise and fall as the armature is turned to different positions. They are transmitted separately to another synchro, and applied to its corresponding stator windings. The composite magnetic field generated by these windings interacts with that generated by the second synchro's rotor winding, which is also excited by 60 or 400 hertz AC, and the pushing and pulling of the lines of force turn the second synchro's rotor to the same position as that on the first synchro. If the first synchro is rotated to a new position, the second takes up the same position also.

Synchro Schematics

The different types of synchros used in synchro systems are listed at the top of the next page.

Those in the first group listed are constructed as shown in Figure 11.1(a), with symbol as illustrated at (b). Transmitters and transformers also used to be known as *generators* or *masters*, and receivers as *motors* or *slaves*. These terms may still be found in some schematics.

The simplest synchro system has a transmitter and receiver connected as in Figure 11.2(a). Whatever position is set on the transmitter is assumed by the receiver. The receiver also rotates in the

Synchro Type	Identification Letters
Control Transformer	CT
Torque Receiver	TR
Control Transmitter	CX
Torque Transmitter	TX
Torque Differential Receiver	TDR
Control Differential Transmitter	CDX
Torque Differential Transmitter	TDX
Resolver	RS

same direction as the transmitter. If the wires are crossed as in Figure 11.2(b) the receiver turns in the opposite direction to the transmitter.

The second group of synchros has three rotor windings. Figure 11.1(c) shows that in the differential type the armature is wound in the same way as the stator. This allows it to receive two inputs and combine them algebraically. The symbol is shown in Figure 11.1(d). In Figure 11.3(a) a torque transmitter (T) signals the position of its rotor to the stator of the torque differential transmitter (D). Voltages are induced in the three rotor windings of D which are a combination of those transmitted from T and those resulting from the position of D's rotor. The voltages transmitted to R will result in the receiver's rotor assuming a position which is equal to the *difference* between the angular positions of T and D, or R = T−D.

In Figure 11.3(b) the leads have been crossed. In this configuration the position assumed by the receiver will be equal to the *sum* of the angular positions of T and D, or R = T+D.

In Figure 11.3(c) the differential synchro is the *receiver*. Its rotor will assume a position equal to the sum of the angular positions of both transmitters.

The difference between control and torque transmitters and receivers is that control types have more turns and a higher impedance. This improves accuracy, but because current is minimal there is insufficient power output to actuate a control. An amplifier is needed to introduce power into the circuit. Accuracy is also improved by connecting a triple capacitor across the stator leads of a control transformer, as shown in Figure 11.4, to cancel stray magnetizing currents in the stator windings. In this figure (a) is the electronic schematic and (b) is the mechanical symbol.

Resolver

The synchro transmitters we have been discussing operate by generating voltages in their three stator windings that vary according

Synchro Systems

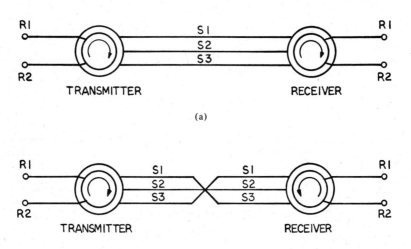

Figure 11.2 Synchro System

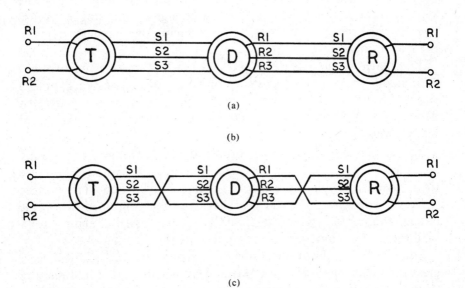

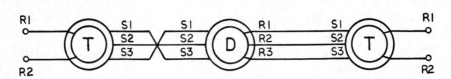

Figure 11.3 Differential Synchro System

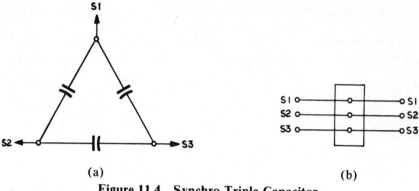

Figure 11.4 Synchro Triple Capacitor

to the angle of the rotor. If the rotor is turned through 360 degrees each stator voltage goes through one cycle of a sine wave with a maximum voltage of 90 volts, where the rotor excitation voltage is 115 volts. However, as the three stator windings are spread equally around the synchro casing the phases of the three signals are staggered 120 degrees. The receiver operates by reproducing the magnetic fields of the transmitter.

The synchro thus has three poles. A resolver is like a synchro, but with many more poles. As its rotor turns and passes each pole a sine-wave voltage is induced. The number of sine waves generated depends upon the number of poles passed, or the angular displacement of the rotor.

To get better resolution the rotor and stator windings are usually arranged so that signals of opposite phase are generated. These are then passed through two clippers or shapers which convert the sine waves to pulses (see chapter 8). They are then combined in a logic circuit (chapter 9) in which the final signal becomes a digital input for a readout device, controller or computer, as the case may be.

Some resolvers, known as optical-electronic resolvers (OERs) use a glass disk with opaque "teeth" printed round the edge. Light passing through the disk and striking a photocell is interrupted by the teeth as the disk turns, producing a train of pulses which are counted by the computer and indicate the angle of rotation. Accuracy is increased by having two outputs of opposite phase, as in the case of electro-mechanical resolvers.

The guidance systems of missiles, space vehicles, aircraft and ships all depend upon the use of gyroscopes ("gyros") and accelerometers

Synchro Systems

(velocity meters). Resolvers "read" their positions and convert the information to digital form, which can then be integrated mathematically by a computer, which in turn issues instructions for adjustments to controls to keep the vehicle on its predetermined course. This same information may also be translated into "human language" for the use of the crew or ground control.

See Figure 11.1(e) for the schematic symbol for a resolver.

Amplidyne

As already mentioned, synchros do not develop any power, and so cannot exert torque to any extent. A device that does develop considerable power is an *amplidyne*, or rotary amplifier. This device steps up the input so that the output will be capable of handling a load.

In Figure 11.5 the symbol for an amplidyne, or regulating generator as it is sometimes called, is shown. The circle represents the armature, and the line crossing it is a connection between two of the four commutator brushes which shorts out the armature winding between them. This allows the creation of strong magnetic fields in the armature, inducing a powerful output current at the other two brushes.

Three of the windings are field coils, and the other two in series with the armature are compensating field windings, which prevent loss of power by canceling opposing fields.

The amplidyne is really a generator in which weak input current changes in the field coils (from a synchro, for instance) are converted to powerful output current changes in the armature by a kind of transformer action known as armature reaction.

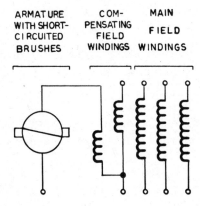

Figure 11.5 Amplidyne (Rotary Amplifier)

12
OTHER CIRCUITS

This chapter could well have been titled "miscellaneous," for in it we deal with circuits which do not fit into any of the categories covered in previous chapters.

MULTIVIBRATORS

A multivibrator is a circuit in which a pair of vacuum-tube or transistor switches turn each other alternately on and off. There are three types:

Astable multivibrators are R-C relaxation oscillators and, as such, are discussed in chapter 4. They may also be called *free-running multivibrators*.

Monostable multivibrators are the electronic equivalent of pushbutton switches. As soon as the input signal is removed they return to their normal state. They are also known as *single-shot multivibrators*.

Bistable multivibrators are the electronic equivalent of double-throw switches. They require a second input signal to return them to their first state. Bistable multivibrators are also called *flip-flops*, or *binary counters*.

MONOSTABLE MULTIVIBRATORS

The monostable or single-shot multivibrator will be considered first. With this type an output is produced whenever the proper input signal is received.

Distinguishing Features

Figure 12.1 is the same plate-coupled free-running multivibrator that was illustrated in Figure 4.12. Figure 12.2 is a single-shot multivibrator, also employing a dual-triode tube.

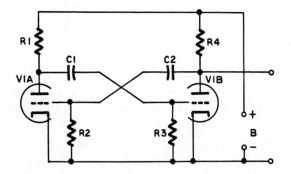

Figure 12.1   Astable or Free-Running Multivibrator

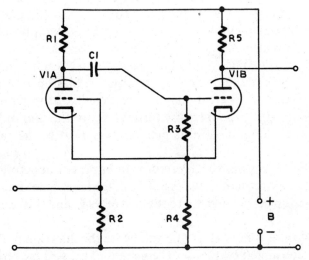

Figure 12.2   Monostable or Single-Shot Multivibrator

Note that the only important difference between them is the absence of the feedback connection in the single-shot version. In Figure 12.1 this connects the plate of V1B via C2 to the grid of V1A.

It is true that the single-shot multivibrator is shown with an input connected across R2, while the free-running multivibrator does not have one, but as you can see by looking at Figure 4.14 the oscillator type may also have an input.

In Figure 12.2 a cathode resistor has also been added.

Use

This circuit is designed to produce an output pulse with a specified duration and amplitude from an input of any shape, provided it has the proper polarity and voltage to trigger the device.

Detailed Analysis

DC Subcircuit: In the absence of any input signal, electron flow is from B− via R4 to both cathodes, and from V1A's plate via R1, and V1B's plate via R5, back to B+. Since R3 connects the grid of V1B to the upper end of R4 the grid is not biased by the voltage drop across R4, and consequently the grid voltage is essentially the same as the cathode voltage. Therefore V1B is conducting, and its plate voltage is lower than the supply voltage by the value dropped across R5.

However, the grid of V1A is connected to B− via R2, therefore this tube is biased negatively by the voltage drop across R4, and is cut off. The plate voltage of V1A is therefore equal to the supply voltage, and capacitor C1 is also charged to the same value on the side toward V1A.

AC Subcircuit: When a positive trigger pulse is applied across R2 of sufficient amplitude to overcome the negative bias on V1A and allow the tube to conduct, there is an immediate voltage drop across R1, and this sends a negative-going pulse through C1 to the grid of V1B. V1B now ceases to conduct, and its plate voltage rises to the value of the supply voltage.

After the trigger pulse, V1A returns to its non-conducting state and its plate voltage returns to the supply voltage level. The rise in positive voltage is coupled through C1 to the grid of V1B, and this tube resumes conduction.

The length of time V1B is off, and hence the duration of the output pulse, is determined by the R−C constant of C1 and the resistors R1, R3 and R4, in series across the supply voltage.

SCHMITT TRIGGER

The Schmitt trigger (Figure 12.3) is a variation of the preceding circuit, and illustrates a typical transistor single-shot multivibrator. Like the vacuum-tube version, it requires an input trigger of proper polarity and voltage, and produces an output of fixed amplitude and duration, determined by the circuit constants. R2 is connected in parallel with C1 to complete a voltage divider, R1, R2 and R5, that stabilizes the base voltage of Q2, which is normally conducting, since its base is more positive than its emitter. Q1 is normally not conducting since the top of R4, though less positive than Q2's base, makes Q1's emitter positive with respect to Q1's base, which is connected to zero potential via R3. The operation of the circuit when a positive input pulse turns Q1 on is the same as for the vacuum-tube circuit.

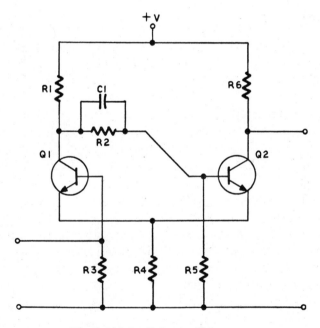

Figure 12.3   Schmitt Trigger

BISTABLE MULTIVIBRATOR, OR FLIP-FLOP

Figure 12.4 shows an Eccles-Jordan flip-flop circuit which is typical of bistable multivibrators.

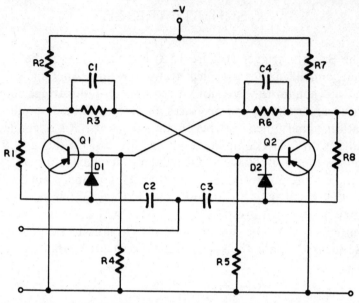

Figure 12.4   Bistable Multivibrator, or Flip-Flop

Distinguishing Features

In this circuit it appears as if the feedback connection of Figure 12.1, removed in Figure 12.2, has been restored. However, the presence of the two diodes indicates that this circuit is not really a free-running multivibrator despite appearances. These diodes form a *steering circuit* that directs the input to each transistor in turn, so we have actually two single-shot multivibrators. If Q1 is off and Q2 on, the input goes to Q1, which then turns on while Q2 turns off. The next pulse is then steered to Q2, turning it back on, and Q2 is turned off again.

Uses

Counters, computers and industrial controls.

Detailed Analysis

Q1's collector is connected to a voltage divider consisting of R2, R3

Other Circuits

and R5. Q2's collector is connected to one consisting of R7, R6 and R4. When Q1 is turned off the supply voltage is divided between R2, R3 and R5, so that the collector of Q1 and the base of Q2 are both negative with respect to the emitters. Q2 therefore has forward bias and turns on; consequently the voltage on the collector is zero, while Q1, with zero bias, remains turned off.

Diode D1 also has zero voltage on the cathode, but a negative voltage reaches its anode via R1. D2 has a negative voltage on its cathode but no voltage on its anode, which is connected to Q2's collector via R8. Consequently D1 is turned off but D2 is on.

If a positive pulse appears at the input it goes to both diodes, but as only D2 is conducting, it only reaches Q2's base. Here it overcomes the negative bias on the base, and causes Q2 to turn off. As a result its collector voltage rises, and the negative-going change in the voltage is coupled through to Q1's base, turning Q1 on. Q1's collector voltage drops to zero, carrying Q2's base voltage to zero also, so the two transistors are now in exactly the opposite state from when the input pulse arrived. Since Q2's collector voltage has now risen from zero to a substantial proportion of the supply voltage, a negative voltage now appears at the output.

The bias on the two diodes is now the opposite of what it was before. Consequently, the next trigger pulse will be steered to Q1's base. Since Q1 is now in the identical state that Q2 was in, while Q2 is now in Q1's former state, Q1 now turns off, turning Q2 on again, and Q2's collector voltage now becomes zero once more.

The speed at which this circuit can switch depends upon the values of R1—C2 and R8—C3. Both resistors and capacitors cannot be very small values, or the input pulse will not be able to switch the transistor, but will appear at the output. Speed can be improved by connecting diodes parallel with R1 and R8.

Another method is to remove R1—C2, R8—C3 and both D1 and D2. The emitters are connected together, and a resistor placed between them and the low side of the circuit (like R4 in Figure 12.3). The input pulse is then applied across this resistor. It momentarily interrupts the current flow in the "on" transistor by reversing the bias. This type of bistable multivibrator is called *emitter-triggered*. Another type is *collector-triggered*. However, the Eccles-Jordan type was the most widely used before the evolution of microelectronic devices necessitated designing flip-flops without capacitors (see Figure 9.13 and corresponding text).

SAWTOOTH GENERATOR

In chapter 4 the use of relaxation oscillators for the generation of sawtooth waveforms was discussed. However, certain applications require a more linear sawtooth than these can produce. Figure 12.5 shows such a type.

Distinguishing Features

R2, R3 and C1 form a timing circuit in which R3 and C1 are usually two of a set of resistors and capacitors that can be switched in and out of the circuit to obtain different sawtooth slope speeds. R2 is a calibration adjustment. The circuit may employ all tubes or all transistors, or both, as shown.

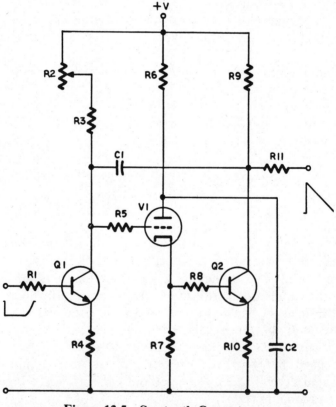

Figure 12.5 Sawtooth Generator

Other Circuits

Uses

Mostly used in oscilloscope sweep generators.

Detailed Analysis

The input to this circuit usually comes from a single-shot multivibrator, such as a Schmitt trigger, so that each sawtooth wave is triggered reliably, not from random pulses or noise spikes.

The proper pulse arrives on the base of Q1, which is called the *disconnect amplifier* (sometimes a diode is used). This transistor is normally conducting, and current flows up the path R4, Q1, R3 and R2 to the positive side of the supply voltage. This voltage divider maintains a constant voltage on the grid of V1 which is slightly negative relative to its cathode, but not enough to cut it off.

When Q1 switches off, this situation changes. The voltage on Q1's collector now starts to rise toward the value of the supply voltage. The rate of this rise is controlled by the R3—C1 combination, since the capacitor can only charge at the R—C time-constant of the two.

As the voltage rises the tube's conduction increases. The current through R7 increases, and the voltage drop across it increases. Consequently the base of Q2 goes more positive.

This results in Q2's conductance increasing in similar proportion, so that its collector voltage goes down.

This negative-going voltage is fed back to the grid of V1 via C1, and tends to oppose the rising positive grid voltage, thus maintaining a linear rise in this voltage, so that a linear output waveform is obtained.

A feedback from the output to the Schmitt trigger (not shown) turns it off at the end of the ramp, and it turns Q1 back on by removal of the pulse on its base. This places the circuit in readiness for the next trigger pulse to start the next sweep.

FREQUENCY DIVIDER

Figure 12.6 illustrates a typical frequency divider, in which an output pulse is produced only after a certain number of input pulses have been received. For example, if it takes six input pulses to produce one output pulse, the output frequency is only one-sixth of the input frequency.

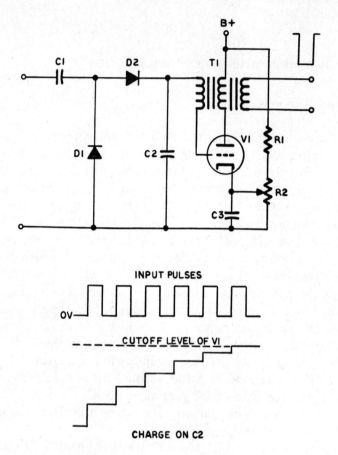

Figure 12.6   Frequency Divider

Distinguishing Features

In this circuit the three-winding transformer connected between the plate and grid of V1 is distinctive, especially when taken in conjunction with the two diodes in the input and the positive voltage applied to the cathode from the variable resistor R2, which, with R1, forms a voltage divider between B+ and the low side of the circuit.

Uses

Counters and similar applications

Detailed Analysis

V1 is biased beyond cutoff by the positive voltage on the cathode, which can be adjusted by R2.

When a positive-going pulse arrives at the input it is coupled through C1 and D2 to C2. Any negative-going portions of the pulse are bypassed by D1. C1 charges up slightly, but cannot charge more because the pulse does not last long enough. However, the charge cannot leak away because D2 cannot conduct in the other direction, and V1 is cut off.

Successive pulses gradually build up the charge on C2 until the voltage on V1's grid reaches the level where the tube can conduct. V1 becomes a low impedance, discharging C2 and dropping the plate voltage. As soon as C2 discharges, however, the tube cuts off again, the plate voltage rises to its former value, and the cycle is repeated.

Circuit Variations

Instead of a three-winding transformer, one with two windings only may be used in conjunction with a dual-triode tube in which one triode replaces the third winding. Both grids and cathodes are connected together. When the first triode conducts, the second does so also, producing an output pulse in the same way as the third winding of the transformer in the previous example.

Thyratrons or silicon-controlled rectifiers may also be used instead of triodes, but since they do not stop conducting under the same conditions as a tube, slightly different circuit arrangements will be required (see chapter 7).

FREQUENCY MULTIPLICATION

The broadcast frequency of a radio station may be many times that of the master oscillator, as we saw in chapter 4. The carrier frequency may be doubled by using the circuit of Figure 12.7. Several stages in succession ("cascaded") will multiply it as many times as necessary.

Distinguishing Features

This circuit is basically the same as that of Figure 3.10, which was a

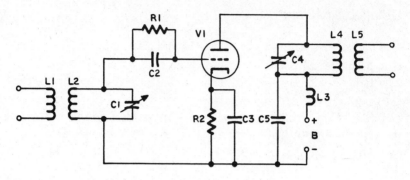

Figure 12.7   Frequency Doubler

Class C RF power amplifier. The only real difference is that there is no neutralizing capacitor in Figure 12.7.

Uses

Transmitters

Detailed Circuit Analysis

This RF power amplifier was discussed in chapter 3, so we need not go over its operation as an amplifier again. The input resonant circuit L2—C1 is tuned to the frequency of the signal from the previous stage.

The output tank circuit L3-C4 is tuned to double this frequency. The tube generates harmonics of the fundamental frequency, and the second, which is double the fundamental, makes the output tank oscillate. In a circuit where the output and the input are at the same frequency this harmonic will be much attenuated, because the output tank is not tuned to that frequency.

Since the output frequency is not the same as the input, oscillation caused by feedback is not as much of a problem, and feedback neutralization is not usually required.

FREQUENCY COMPARATOR

The frequency comparator circuit in Figure 12.8 is similar to the discriminator circuit of Figure 6.7 used in FM receivers.

Other Circuits

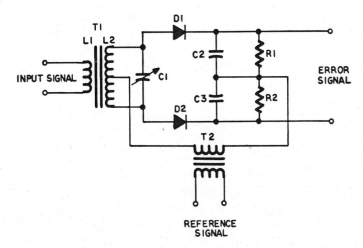

Figure 12.8   Frequency Comparator

Distinguishing Features

This circuit only differs from a discriminator in having T2 separate, and not a part of T1.

Uses

Use of a discriminator to control frequency was mentioned in connection with FM modulation in chapter 5. Such a discriminator operates as a frequency comparator. The circuit may be used in any application where automatic frequency control is required.

Detailed Analysis

A control or standard frequency from a crystal-controlled oscillator is coupled into the circuit through T2. A sample of the signal to be monitored is applied across T1.

The operation of the circuit is the same as the operation of the discriminator (see chapter 6). As long as no difference exists between the input signal and the reference signal the voltages at each end of L2 remain equal, consequently the charges on C2 and C3 are equal also. Since they cancel each other no output voltage is produced.

When the input signal changes so that it is no longer at the resonant frequency of L2—C1 the two halves of L2 develop different voltages due to the resultant phase shifts. C2 and C3 are now charged to different voltages, so an error voltage appears at the output. This error voltage may be used to correct the signal frequency by using a *reactance circuit,* as explained in chapter 5.

PHOTOCELLS

Transducers are devices for converting physical quantities of one kind into corresponding physical quantities of another kind. A loudspeaker changes electrical signals into sound waves, a microphone changes sound waves into electrical signals.

Transducers which do the same for light are also of two kinds:

1. *Photosensitive,* changing light energy to electrical energy.
2. *Photoemissive,* changing electrical energy to light energy.

Photosensitive devices are also of two main types.

1. *Photovoltaic,* generating a voltage; and
2. *Photoconductive,* allowing current to pass.

Figure 12.9 illustrates a photovoltaic transducer or photocell (B) connected in a circuit in which the voltage produced by light falling on it controls a transistor, which amplifies it. Note that the symbol for the photocell is that of a voltaic cell enclosed in a circle, with arrows pointing towards it to represent incident light. The older device was a phototube, and a similar vacuum-tube circuit was used in automobiles for automatic dipping of the headlights. A rather more elaborate circuit is used for "reading" the sound track on the edge of a motion-picture film and converting it to an audio signal.

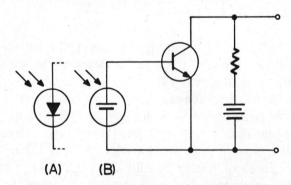

(A)   (B)

Figure 12.9 Photovoltaic Transducers

The solar cell used in solar batteries is also a photovoltaic device, though the idea here is simply to generate power, not to drive an amplifier.

The other device, illustrated at (A), is a photodiode, which changes from non-conduction to conduction under the influence of light.

Figure 12.10 illustrates a *photoconductive* device (a cadmium-sulfide photocell) as used in a modern camera. In this circuit the resistance of the cell varies with the intensity of light falling on it (reflected from the scene viewed by the camera), so that the reading of the exposure meter E, which depends on the battery current passed by the cell, indicates the light value. A variable resistor allows the current to be adjusted to bring the meter pointer or other indicating device back to midscale or some other index, to compensate for different film speeds, lens apertures, and so on. In some cameras this is done automatically.

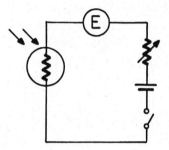

Figure 12.10   Photoconductive Transducer

Lamps of all kinds are the most well-known *photoemissive* devices, but a more recent development is that of *light-emitting diodes* (LEDs), which because of their small size and low current consumption can replace lamps with advantage in many applications, especially as indicators and in readout displays. It is possible that they may also one day be used for TV screens.

CATHODE-FOLLOWER CIRCUIT

Figure 12.11 illustrates a cathode-follower circuit, which is a special type of amplifier circuit having a voltage gain of less than unity, though there may be a power gain.

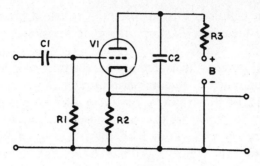

Figure 12.11   Cathode Follower

Distinguishing Features

The cathode-follower circuit has the same features as the voltage amplifiers of chapter 2, except that:

1. The plate is grounded for signals through C2.
2. The output is taken from across R2.
3. There is no cathode-bypass capacitor

Uses

To achieve uniform response over a wide range of frequencies, an amplifier should have a low input capacitance and a low load impedance. The frequency response may also be improved by degenerative feedback, as we saw in chapters 2 and 3. The cathode follower, or emitter follower in the case of transistor circuits, possesses these qualities, and in addition it may be used to match the impedance of one circuit to that of another. It is used, therefore, as a circuit intermediate between others. For example, a conventional amplifier with a high output impedance requires a cathode follower between its output and a transmission line with a low impedance. A coaxial transmission line typically has an impedance of 50 or 75 ohms, so power would be lost and standing waves created by the mismatch if a cathode-follower circuit was not employed.

Detailed Analysis

DC Subcircuit: Electron flow is from B– via R2, V1 and R3 back to B+. R2 provides grid bias by making the cathode more positive than the grid because of the voltage drop across the resistor, and R3 drops the B+ voltage to the correct operating value for the plate.

AC Input Circuit: Input signals are applied between grid and cathode of V1 as in other amplifiers, and using standard types of coupling from the previous circuit. This example is RC-coupled via C1 and R1.

AC Output Circuit: Variations in signal voltage on the grid of V1 cause corresponding variations in the resistance of the tube and, therefore, in the current flowing through it and through R2. Since R2 is not bypassed an output signal voltage appears across it. No output signal appears across R3 because it is bypassed by C2. The output signal is in phase with the input, which it is therefore said to "follow." However, the absence of a cathode-bypass capacitor introduces *degeneration* (see chapter 2), which is why there is a voltage loss with this circuit instead of amplification.

Circuit Variations

Figure 12.12 gives the transistor equivalent of Figure 12.11. This circuit is called an emitter-follower or *grounded-collector circuit*, because the collector of Q1 is bypassed by C1, so that no signal voltage develops across R1, and the output is taken from R2. The output signal is in phase with the input, and the emitter resistor is not bypassed.

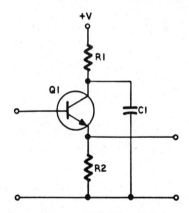

Figure 12.12   Emitter Follower

INHIBITING CIRCUIT

Figure 12.13 shows an inhibiting circuit. The inhibiting signal cuts off the other signal by canceling part of it. T1 is a phase-reversing

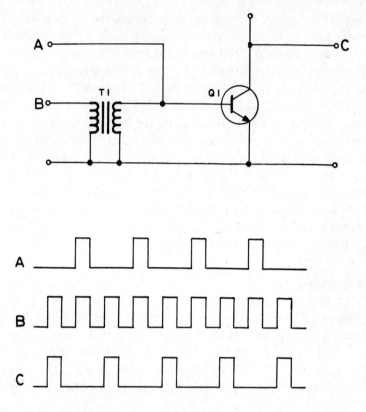

Figure 12.13 Inhibiting Circuit, or Gate

transformer. If the two signals at A and B have the same phase, they will be of opposite phase on the base of Q1. If they are of the same amplitude, cancellation will take place as shown. The residual signal is reversed again to appear at C with the same phase as at B. An inverter could also be used instead of T1 (see chapter 9).

APPENDIX

Amplifier Classification

Amplifiers are classified according to the operating voltages which determine the bias between cathode and grid in the case of vacuum tubes, or across the emitter-base and base-collector junctions in the case of transistors. The relationship between this *bias voltage* and the *cutoff voltage*, when no signal is present, places the amplifier in one of the following classes.

Class A operation requires a bias voltage midway between the cutoff voltage and the bias voltage that would give maximum output current (see Figure A.1). The input signal must be small enough so that its positive and negative excursions away from the resting voltage do not go into the non-linear border areas close to the zero and maximum current levels. The output current will then be a faithful replica of the complete input signal.

Class A operation is therefore suitable for voltage amplifiers, where input signals are small, and faithful reproduction using one tube or transistor is required. Because a high-value load resistor is used to change output current to output voltage the output current cannot be high. This is important, because in Class A operation current flows at all times, with or without an input signal. This means that power is being wasted, and efficiency (ratio of useful output to total power consumed) is only 20-25%.

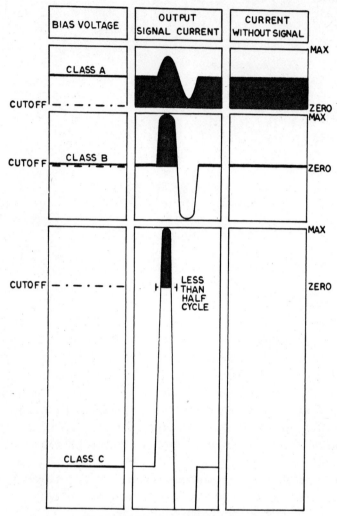

Figure A-1 Amplifier Operation

Class A amplifiers can be single-ended or pushpull, and can be used for both AF and RF.

Class A B_1 operation requires a bias voltage somewhere between the cutoff voltage and the bias required for Class A. A somewhat larger input signal is required but its positive excursion, in the case of a vacuum tube, must not be large enough to draw grid current. Since the opposite excursion of the input signal goes a little beyond cutoff, output current drops to zero, so that the tip of the signal is clipped.

This results in some distortion, and it is preferable for this amplifier to be a pushpull type, so that the other tube or transistor can supply the missing portion. As a larger input signal is required, this amplifier is more suitable for a power output stage. It can be used for both RF and AF.

Class AB_1 amplifier efficiency is better (25-35%) than Class A because less current is wasted in non-productive operation.

Class AB_2 operation is similar to Class AB_1, except that in vacuum tubes the input signal is allowed to be large enough to draw grid current on a positive excursion. The efficiency (35-50%) is better than Class AB_1.

This is always a pushpull amplifier, and may be used for either RF or AF.

Class B operation requires a bias voltage at cutoff. This will give output-current flow only on one half-cycle of the input signal, as shown in Figure A.1; therefore for audio it must always be used in a pushpull circuit. For RF it can be used with a single-ended circuit.

Since current in the absence of an input signal is negligible, efficiency is much higher (60-70%), which is very important in higher-powered amplifiers.

Class C operation requires a bias voltage which may be two or three times the value of the cutoff voltage. Output current can flow only during a portion of the alternate swings of the input signal, which therefore must be very large. However, as this portion is considerably less than a half-cycle the efficiency of a Class C amplifier is higher than any other (it may be over 90%).

Class C amplifiers cannot be used for audio, and are mainly used for RF power stages in a transmitter. The output current flows in a series of powerful pulses that shock-excite the tank circuit into oscillation. As this is an LC resonant circuit its output waveform is a sine wave, varying in amplitude if modulated. Since the resonant circuit impedance is very high compared to that of the tube virtually all the output voltage appears across it.

Class C amplifiers can be either pushpull or single-ended. Because they are operated intermittently RF output power-amplifier tubes can be driven very much harder than if they had to run continuously.

Transistor Operation

You can think of a transistor as consisting of a pair of diodes joined back to back. A diode, whether germanium or silicon, comprises P-

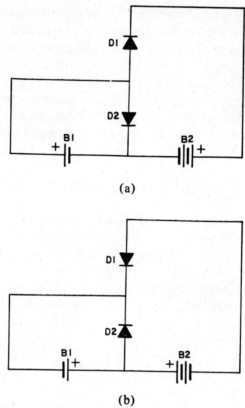

Figure A-2 Transistor Operation

type and N-type "regions" which meet at a *junction*. To make a diode conduct, it must be *forward-biased*, which means that an external voltage has to be applied so that the *P* region is *P*ositive with respect to the *N* region which, conversely, is *N*egative. The P region is the anode, and the N region is the cathode.

In Figure A.2 an NPN transistor (a) and a PNP transistor (b) are drawn as if they were pairs of diodes. In both transistors D2 is forward-biased by B1, and D1 is reverse-biased by B2, consequently D2 should conduct and D1 should not.

Forward-biasing allows conduction by lowering the *barrier voltage* at the junction. Conversely, reverse-biasing raises it. However, because of the proximity of the junctions in a transistor, D1 is influenced by D2. Therefore when D2 is forward-biased both diodes conduct, and when D2 is reverse-biased neither does. Actually, the purpose of D1 is

Appendix 281

to increase the flow of current from B2 through D1 and D2, so that amplification can take place, since a single diode cannot amplify.

In real circuits B1 is replaced by a voltage divider that drops B2's voltage to the proper value. The forward-bias voltage required to overcome the barrier at the emitter-base junction (D2 in Figure A.2) is .2 volt for germanium and .7 volt for silicon.

When a signal voltage is applied across the emitter-base junction it raises and lowers the barrier voltage in accordance with its own voltage fluctuations, thereby varying the current flow through the transistor (D1 + D2).

The *MOS/FET* ("metal-oxide semiconductor/field-effect transistor") illustrated in Figure A.3 is made by oxidizing the surface of a thin wafer of N-type silicon called the substrate. This oxide, which is a type of glass, is then removed from the source and drain areas by photo-etching. The wafer is then placed in a furnace in which a P-type impurity (boron, aluminum or gallium) is present as a gas that diffuses into the exposed areas, converting them to P-type silicon. A similar process is used to deposit thin films of aluminum to form metal

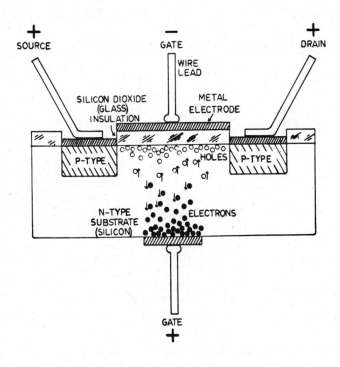

Figure A-3   MOS/FET

electrodes, and then gold wire leads are bonded to them by a combination of heat and pressure.

In this transistor the *emitter, base* and *collector* have been renamed *source, gate* and *drain*. A fourth connection is often made to the substrate, as shown.

Electrons are the majority carriers in n-type silicon, but if a sufficiently negative voltage is applied to the gate electrode, the resulting field repels them from the vicinity of the silicon-silicon dioxide interface, causing a *depletion layer* populated only by positive minority carriers, hence the term *p-channel* for this type of MOS/FET. This layer is very thin, its depth varying with the electric field strength, which in turn depends upon the gate potential. Conduction increases with field intensification, hence the term *enhancement type*. No conduction can take place without a gate potential. Another device, which works in the opposite way, is called a *depletion type*. In it, current flows at all times except when a large enough potential on the gate cuts it off entirely.

MOS/FET's are used in integrated circuits, where they are preferred because it costs less to make an IC with MOS/FET's than with junction transistors. However, they are slower than the latter, so they are mostly used where high-speed operation is not required.

Figure A.4 shows a MOS/FET amplifier, in which Q1 acts as the

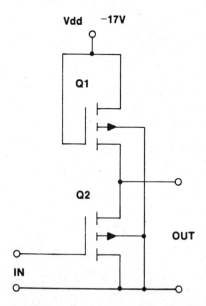

Figure A-4 MOS/FET Amplifier (Driver-Load Pair)

Appendix

load resistor. As explained in chapter 2, transistors are often used with a fixed bias to provide resistance values higher than could be economically provided in an IC.

Color Coding

Fixed non-metallic resistors and some other components are color-coded to indicate their values instead of having the information printed or stamped on them. The colors used and their numerical significance are as follows:

Color	Value
Black	0
Brown	1
Red	2
Orange	3
Yellow	4
Green	5
Blue	6
Violet	7
Gray	8
White	9

On a *composition resistor* there are three or four colored bands around the resistor body, adjacent to one end. Starting with the one nearest the end, the first two colors are the significant figures of the resistor value in ohms, and the third is the power of ten of the multiplier. For example, blue (6), red (2) and yellow (4) should be read as 62×10^4, or 620 kilohms. If the third band is gray (or silver), or white (or gold), the multiplier is 10^{-2}, or 10^{-1}, respectively.

If there is no fourth band the tolerance is ±20%. Otherwise, silver denotes ±10%, gold ±5%. Other tolerances may be indicated by using the applicable colors, but these are seldom encountered as most precision resistors have their values stamped or printed on them.

Some *film-type* resistors also may be color-coded, in which case an additional color may be added at the front end, to provide for three significant figures, followed by a multiplier and tolerance.

Capacitors, for the most part, are not color-coded. Exceptions are some *ceramic* and *mica* types. Ceramic *tubular* capacitors may be coded in the same way as composition resistors, except that an additional color is added in front of the others to indicate the *temperature characteristic.* The colored bands may also be replaced by colored dots. Ceramic *disks* have these arranged around the edge, to be read clockwise. Some very small tubular capacitors have only three dots, which give two significant figures and the multiplier only.

Most *mica* capacitors are no longer color-coded. However, the flat rectangular molded silvered-mica type may employ an arrangement of six colored dots arranged three on each side of an arrow, or some other mark to indicate which way to read them. With the arrow pointing to the right the upper three dots (reading from left to right) indicate EIA standard (this is a white dot, which may be omitted) and two significant figures. The bottom row (reading from right to left) gives the multiplier, tolerance and type.

The significant figures always give the value in picofarads. The temperature characteristic (ceramic capacitors) or type (mica capacitors) is denoted as follows:

COLOR	TEMPERATURE CHARACTERISTIC	TYPE (MFR'S SPEC.)
Black	NP0	A
Brown	N033	B
Red	N075	C
Orange	N150	D
Yellow	N220	E
Green	N330	
Blue	N470	
Violet	N750	

Diodes used in signal circuits are often color-coded because of their small size. Diode designations are always given, for example, in the form 1N914A. The 1N- is common to all and does not have to be indicated, therefore the color bands denote the figures and letter following. Suffix letters are coded as follows:

Black	(no suffix)
Brown	A
Red	B
Orange	C
Yellow	D
Green	E
Blue	F
Violet	G
Gray	H
White	J

The colored bands are grouped at the cathode end of the diode.

Transformer leads are color-coded as follows:

Appendix

POWER TRANSFORMERS

Primary (tapped)

Black (common)
Black/yellow
  (centertap)
Black/red

or
Primary (untapped)
Two black leads

Secondary

High voltage   { Red / Red/yellow (centertap) / Red

Rectifier filament   { Yellow / Yellow/blue (centertap) / Yellow

Amplifier filament #1   { Green / Green/yellow (centertap) / Green

Amplifier filament #2   { Brown / Brown/yellow (centertap) / Brown

Amplifier filament #3   { Slate / Slate/yellow (centertap) / Slate

IF TRANSFORMERS

Primary
Blue (plate)

Red (B+)

Secondary
Green (grid or diode)
Green/black (full-wave diode)
Black (grid or diode return, AVC or ground)

AUDIO & OUTPUT TRANSFORMERS

Primary
Blue (plate)
Red (B+)
*Blue or brown (plate)

Secondary
Green (grid or voice coil)
Black (return or voice coil)
*Green or yellow (grid)

*Pushpull only.

Preferred Values

Most resistors and capacitors other than precision or special-buy items are supplied in *preferred values,* to simplify manufacture and stocking by limiting the number of different values to bare essentials. The tolerance provides the necessary spread to cover the gaps between the nominal values. Preferred values for the three principal tolerances are manufactured for each multiplier as follows:

20%	10%	5%
10	10	10
	12	12
15	15	15
		16
	18	18
		20
22	22	22
		24
	27	27
		30
33	33	33
		36
	39	39
		43
47	47	47
		51
	56	56
		62
68	68	68
		75
	82	82
		91
100	100	100

International System of Units (SI Units)

SI is a simplified system in that it is founded upon seven units of measure called base units. Two additional units, generally referred to as supplementary units, complete the foundation from which units for measuring other physical quantities are derived mathematically.

SI BASE UNITS

Unit Name	Plural Form	Pronunciation	Symbol	Quantity
ampere	amperes	am'pi(ə)r	A	electric current
candela	candelas	kăn de'lə	cd	luminous intensity
kelvin	kelvins	kĕl'v ᵊn	K	thermodynamic temperature*
kilogram	kilograms	kĭlō'grăm	kg	mass
metre**	metres	mē't ᵊr	m	length
mole	moles	mōl	mol	amount of substance
second	seconds	sek'ənd	s	time

*Degree Celcius (°C) accepted (°C =K 273.15). Plural form is degrees Celsius.
**The spelling "meter" is also acceptable.

SI SUPPLEMENTARY UNITS

Name	Plural	Pronunciation	Symbol	Quantity
radian*	radian	rā'dē ən	rad	plane angle
steradian	steradian	stə rā'dē ən	sr	solid angle

*Use of degree, minute, and second is acceptable.

SI DERIVED UNITS WITH SPECIAL NAMES

Unit Name	Plural Form	Pronunciation	Symbol	Quantity	Formula
becquerel	becquerels	bē krel'	Bq	radioactivity	s^{-1}
coulomb	coulombs	kü' lŏm	C	electric charge	A·s
farad	farads	far' ad	F	electric capacitance	C/V
gray	grays	grā	Gy	absorbed dose	J/kg
henry	henries	hen' re	H	inductance	Wb/A
hertz	hertz	hərts	Hz	frequency	1/s or s^{-1}
joule	joules	ju(e)l*	J	energy	N·m
lumen	lumens	lü m' ən	lm	luminous flux	cd·sr
lux	lux	laks'	lx	illuminance	m^{-2}·cd·sr
newton	newtons	n(y)u' tən	N	force or weight	m·kg·s^{-2}
ohm	ohms	om	Ω	electric resistance	V/A
pascal	pascals	pas' kəl	Pa	pressure or stress	N/m^2
siemens	siemens	sē' məns	S	conductance	A/V
tesla	teslas	tes' lă	T	magnetic flux density	Wb/m^2
volt	volts	vōlt'	V	electric potential	W/A
watt	watts	wăt'	W	power	J/s
weber	webers	web' ər	Wb	magnetic flux	V·s

*Usually pronounced joul in U.S.

SOME DERIVED UNITS WITHOUT SPECIAL NAMES

Unit Name	Symbol	Quantity
square metre	m^2	area
cubic metre	m^3	volume
metre per second	m/s	velocity (linear)
radian per second	rad/s	angular velocity
metre per second squared	m/s^2	acceleration (linear)
radian per second squared	rad/s^2	angular acceleration
newton metre	N·m	moment of force (torque)
kilogram per cubic metre	kg/m^3	density
joule per kelvin	J/K	entropy
watt per square metre	W/m^2	thermal flux density

NON-SI units used in specialized fields and those of practical importance will remain in use internationally.

NON-SI UNITS MOST COMMONLY USED WITH SI

Name	Plural	Symbol	Value in SI
minute	minutes	min	1 min = 60s
hour	hours	h	1 h = 60 min = 3,600s
day	days	d	1d = 24 h = 86,400s
degree	degrees	°	1° = $(\pi/180)$ rad
minute	minutes	′	1′ = $(1/60)°$ = $(\pi/10,800)$ rad
second	seconds	″	1″ = $(1/60)′$ = $(\pi/648,000)$ rad
litre	litres	l	1 l = 1 dm^3 = $10^{-3} m^3$
metric ton	metric tons	t	1 t = 10^3 kg
bar	bar or bars	bar	1 bar = 10^5 Pa

Appendix

PREFIXES

Standard prefixes that are used with the above to obtain units of convenient size are as follows:

Name	Pronunciation	Symbols	Amount	Multiples and Submultiples	Definition
exa	ex′a	E	1 000 000 000 000 000 000	10^{18}	one million million million times
peta	pet′a	P	1 000 000 000 000 000	10^{15}	one thousand million million times
tera	ter′a	T	1 000 000 000 000	10^{12}	one million million times
giga	ji′ga	G	1 000 000 000	10^9	one thousand million times
mega	meg′a	M	1 000 000	10^6	one million times
kilo	kil′o	k	1 000	10^3	one thousand times
hecto	hek′to	h*	100	10^2	one hundred times
deka	dek′a	da*	10	10	ten times
deci	des′i	d*	0.1	10^{-1}	one tenth of
centi	sen′ti	c*	0.01	10^{-2}	one hundredth of
milli	mil′i	m	0.001	10^{-3}	one thousandth of
micro	mi′kro	μ	0.000 001	10^{-6}	one millionth of
nano	nan′o	n	0.000 000 001	10^{-9}	one thousandth millionth of
pico	pe′co	p	0.000 000 000 001	10^{-12}	one millionth millionth of
femto	fem′to	f	0.000 000 000 000 0001	15^{-15}	one thousandth millionth millionth of
atto	at′to	a	0.000 000 000 000 000 001	10^{-18}	one millionth millionth millionth of

*These prefixes should generally be avoided except for measurement of area and volume, and for nontechnical use of centimeter.

Examples:

Gigahertz (billion hertz) = GHz
Megohm (million ohms) = MΩ
Kilovolt (thousand volts) = kV
Centimeter (hundredth of a meter) = cm
Millisecond (thousandth of a second) = ms
Microhenry (millionth of a henry) = μH
Picofarad (trillionth of a farad) = pF

Conversion Factors (U.S. and Metric)

Ampere-hour	3.6×10^{-3}	coulombs
Ampere-turn	1.257	gilberts
Angstrom	1.00×10^{-10}	meter
British thermal unit (BTU) (mean)	1.05587×10^3	joules
Centimeter	3.937×10^{-1}	inches
Coulomb	2.778×10^{-4}	ampere-hours
Foot	3.048×10^{-1}	meters
Gauss	1.00×10^{-4}	tesla
Gilbert	7.958×10^{-1}	ampere-turns
Grain	6.479891×10^{-5}	kilograms
Gram	3.527×10^{-2}	ounces (avoirdupois)
Horsepower (electric)	7.46×10^2	watts
Inch	2.54	centimeters

Kilogram	2.205	pounds (avoirdupois)
Kilometer	6.2137×10^{-1}	miles
Kilowatt-hour	3.6×10^6	joules
Liter	2.113	pints (US)
Maxwell	1.00×10^{-8}	weber
Meter	1.094	yards
Micron	1.00×10^{-6}	meter
Mile (nautical)	1.852	kilometers
Mile (statute)	1.609344	kilometers
Oersted	7.9577472×10^{-1}	amperes per meter
Ounce (avoirdupois)	28.349523125	grams
Ounce (troy)	31.1034768	grams
Pint (US)	$4.73176473 \times 10^{-1}$	liters
Pound (avoirdupois)	4.5359237×10^{-1}	kilograms
Tesla	1.00×10^4	gauss
Watt	1.341×10^{-3}	horsepower
Weber	1.00×10^8	maxwells
Yard	9.144×10^{-1}	meter

Ohm's Law Formulas

DC:

$E =$	$I =$	$R =$	$P =$
IR	$\dfrac{E}{R}$	$\dfrac{E}{I}$	EI
$\dfrac{P}{I}$	$\dfrac{P}{E}$	$\dfrac{P}{I^2}$	$I^2 R$
$\sqrt{PR}$	$\sqrt{\dfrac{P}{R}}$	$\dfrac{E^2}{P}$	$\dfrac{E^2}{R}$

AC:

$E =$	$I =$	$Z =$	$P =$
IZ	$\dfrac{E}{Z}$	$\dfrac{E}{I}$	$IE\cos\theta$
$\dfrac{P}{I\cos\theta}$	$\dfrac{P}{E\cos\theta}$	$\dfrac{P}{I^2\cos\theta}$	$I^2 Z \cos\theta$
$\sqrt{\dfrac{PZ}{\cos\theta}}$	$\sqrt{\dfrac{P}{Z\cos\theta}}$	$\dfrac{E^2\cos\theta}{P}$	$\dfrac{E^2\cos\theta}{Z}$

Appendix

Common Electrical Formulas

Resistors in series:	$R_T = R_1 = R_2 + \ldots$
Resistors in parallel (two):	$R_T = \dfrac{R_1 R_2}{R_1 + R_2}$
Resistors in parallel (more than two):	$\dfrac{1}{R_T} = \dfrac{1}{R_1} + \dfrac{1}{R_2} + \dfrac{1}{R_3} + \ldots$
Capacitors in series (two):	$C_T = \dfrac{C_1 C_2}{C_1 + C_2}$
Capacitors in series (more than two):	$\dfrac{1}{C_T} = \dfrac{1}{C_1} + \dfrac{1}{C_2} + \dfrac{1}{C_3} + \ldots$
Capacitors in parallel:	$C_T = C_1 + C_2 + \ldots$
Capacitive Reactance:	$X_C = \dfrac{1}{2\pi f C}$
*Inductors in series:	$L_T = L_1 + L_2 + \ldots$
*Inductors in parallel (two):	$L_T = \dfrac{L_1 L_2}{L_1 + L_2}$
*Inductors in parallel (more than two):	$\dfrac{1}{L_T} = \dfrac{1}{L_1} + \dfrac{1}{L_2} + \dfrac{1}{L_3} + \ldots$
Inductive Reactance:	$X_L = 2\pi f L$
Coupled Inductance:	$L_T = L_1 + L_2 \pm 2M$
Transformer ratio	$\dfrac{N_p}{N_s} = \dfrac{E_p}{E_s} = \dfrac{I_s}{I_p} = \sqrt{\dfrac{Z_p}{Z_s}}$
Series impedance:	$Z = \sqrt{R^2 + (X_L - X_C)^2} = \sqrt{R^2 + X^2}$
Parallel impedance:	$Z = \dfrac{RX}{\sqrt{R^2 + X^2}}$
Temperature — Celsius:	$C = 0.556F - 17.8$
Temperature — Fahrenheit:	$F = 1.8C + 32$
Temperature — Kelvin:	$K = C + 273$
Decibels — power:	$dB = 10 \log \dfrac{P_1}{P_2}$

---

*No coupling between coils

Decibels — voltage: $\quad dB = 20 \log \dfrac{E_1}{E_2}$

Power factor: $\quad pf = \dfrac{P}{EI} = \dfrac{R}{Z}$

Resonance: $\quad f = \dfrac{1}{\sqrt{2\pi LC}}$

Greek Alphabet

Name	Upper Case	Lower Case	Used For
Alpha	A	α	Angles, attenuation constant, absorption factor
Beta	B	β	Angles, phase constant
Gamma	Γ	γ	Conductivity, propagation constant, angles (upper or lower case)
Delta	Δ	δ	Changing value, increment or decrement & angles
Epsilon	E	ϵ	Natural logarithm base: 2.71828, dielectric constant
Zeta	Z	ζ	Impedance
Eta	H	η	Efficiency, hysteresis
Theta	Θ	θ	Phase angle, time constant, reluctance
Iota	I	ι	Unit vector
Kappa	K	κ	Coefficient of coupling, susceptibility
Lambda	Λ	λ	Wavelength, attenuation constant
Mu	M	μ	Micro-, amplification factor, permeability
Nu	N	ν	Reluctivity, frequency
Xi	Ξ	ξ	
Omicron	O	o	
Pi	Π	π	3.14159 26535 89793 23846 ...
Rho	P	ρ	Resistivity
Sigma	Σ	σ	Summation (upper case), conductivity, deviation
Tau	T	τ	Time-constant, time-phase displacement, transmission factor
Upsilon	Υ	υ	

Appendix

Phi	Φ	φ	Angles, magnetic flux, scalar potential (upper case)
Chi	X	χ	Susceptibility, angles
Psi	Ψ	ψ	Dielectric flux, phase difference, angles
Omega	Ω	ω	Ohms (upper case), angular velocity (= $2\pi f$)

Note: Lower case letter is used except where upper case is indicated.

Radio Frequency Spectrum

Band Numbers	Frequency Range	Metric Subdivision	Abbreviation	Full Name
2	30-300 Hz	Megametric waves	ELF	Extremely low frequency
3	300-3000 Hz	—	VF	Voice frequency
4	3-30 kHz	Myriametric waves	VLF	Very-low frequency
5	30-300 kHz	Kilometric waves	LF	Low frequency
6	300-3000 kHz	Hectometric waves	MF	Medium frequency
7	3-30 MHz	Decametric waves	HF	High frequency
8	30-300 MHz	Metric waves	VHF	Very-high frequency
9	300-3000 MHz	Decimetric waves	UHF	Ultra-high frequency
10	3-30 GHz	Centimetric waves	SHF	Super-high frequency
11	30-300 GHz	Millimetric waves	EHF	Extremely-high frequency
12	300-3000 GHz	Decimillimetric waves	—	—

Examples of Use

Band Numbers	Service
2-4	(Below 10 KHz) not allocated
4	(Above 10 KHz) radio navigation, maritime mobile and fixed
5	Radio navigation, maritime mobile and fixed
6	Broadcasting (535-1605 KHz), aircraft, distress, amateur
7	Broadcasting (short wave), aircraft, amateur, citizens' band
8	TV, FM, satellites, amateur, space operations, radio location, public safety, aircraft
9	Aircraft, amateur, satellites, space operations, radio navigation, citizens' band TV
10	Radar, satellites, space operations, deep space research, amateur, navigation
11	Radio astronomy, space research, radar, satellites, navigation satellites, amateur
12	(Above 275 GHz) not allocated, amateur

INDEX

A

A (class letter), 18
accelerator grid, 182
active elements, 38, 39, 40, 41
active filters, 229
adjustable capacitor (symbol), 21
adjustable inductor (symbol), 24
AGC, 88, 169
amplidyne, 259
amplifier:
   adjustable gain, with (symbol), 18
   audio voltage, direct coupled, 61
   audio voltage, RC-coupled, 53
   classification, 277
   correction, 148
   current, 96
   general (symbol), ·8
   magnetic, 113
   power, 96
   video, 78
   voltage, 52
amplitude distortion, 103
analog computer, 78
AND gate, 234, 236
antenna(s), 107, 162
antenna (symbol), 18
Armstrong balanced modulator, 147
Armstrong oscillator, 127
astable multivibrator, 132, 260
asymmetrical pads, 221
attenuators, 221
audible signaling device(s) (symbols), 19
automatic gain control (AGC), 169
automatic gain control, keyed, 172
automatic volume control (AVC), 168
autotransformer (symbol), 32

B

B (class letter), 29
balanced filters, 226
balanced pads, 221
balun transformer, 93
bandpass and matrix circuit, 187
bandpass filter, 223
bandstop filter, 224
base, 56
base tone control, 60
battery, 191
battery (symbol), 19
beta (β) multiplier, 66
bistable multivibrator, flip-flop, 244, 263
bits, 246
block diagram, 34
blocking oscillator, 128
breakdown diode (symbol), 26
breakover voltage, 219
bridge circuits, 248
   Hay, 251
   LC, 249
   Maxwell, 251
   Owen, 251
   resonance, 252
   Schering, 252
   Wheatstone, 248
   Wien, 252
bridged H-pad, 223
bridged T-pad, 223
bridge-rectifier power supply, 201
brute force, 49
BT (class letter), 19
buffer capacitor, 218
buzz control, 184
bypass capacitor, 58, 98, 99

C

C (class letter), 21
cable (symbol), 20
capacitance, distributed, 82
capacitor(s):
 buffer, 218
 bypass, 58, 98, 99
 general (symbols), 21
 polarized (symbols), 21
 trimmer, 162
 variable (symbols), 21
cascade amplifier, cascading, 63
cascode amplifier, 91
cathode, 55
cathode feedback, 118
cathode-follower circuit, 273
cathode-ray tube (CRT), 208
CB (class letter), 22
chassis ground (symbol), 23
choke, 54, 195
choke, swinging, 207
chopper, 216
circuit analyzer, 38, 39, 40
circuit breaker (symbol), 22
circuit identification, 38
circuit locator, 41
circuit return (symbol), 23
clamp circuits, 233
class letters, 17
clippers, 231
closed-loop gain, 77
coaxial cable (symbol), 20
collector, 56
collector feedback, 119
color coding, 283
color detector (synchronous detector), 187
color gain control, 188
Colpitts oscillator, 120
common cathode, 54
common emitter, 54, 78, 86
common-mode rejection, 68
comparator, 204
complementary symmetry, 102
conductor (symbol), 20
confirm complaint, 45
connectors (symbols), 22

constant-current source, 71
constant-K filters, 224
contrast control, 82
conversion, converter, 86, 150, 160
conversion factors (US and metric), 289
correction amplifier, 148
coupling, 52
CR (class letters), 26
cross-neutralization, 111
crystal, 126
crystal oscillator, 124, 125
crystal oven, 126
crystal unit, piezoelectric (symbol), 23
current amplifier, 54, 96
current measurement, 50

D

D (class letter), 26
damper tube, 212
Darlington amplifier, 66
DC inverters, 215
DC restoration, 233
DC voltage analysis, 50
decoupling, 82, 85
de-emphasis, 187
degeneration, degenerative feedback, 66, 77, 81
delta modulation, 156
demodulation, demodulator, 166
 diode detector, 166, 170
 grid-leak detector, 174
 regenerative detector, 176
 video detector, 169
diac, 219
diac (symbol), 27
difference signal, 147
differential amplifier, 66, 68
differential comparator, 72
differentiation, 229
dimmer control circuit, 219
"dip-chip," 74
dipole (symbol), 18
direct-coupled audio-voltage amplifiers, 61
disconnect amplifier, 267
discriminator, 146, 178

Index

distortion, amplitude, 103
distortion, frequency, 104
distributed capacitance, 82
DM, 156
double-emitter follower, 66
doubly balanced modulator, 153
DS (class letters), 24
DTL NAND gate, 241
dual-diode-triode tube, 171
dual-in-line-packaged integrated circuit, 74
dual triode, 92
dual triode (symbol), 32
dynamic tests, 46
dynamotor, 216

E

E (class letter), 18
earth ground (symbol), 23
Eccles-Jordan flip-flop, 263
electrical formulas, common, 291
electron-coupled oscillator, 122
electron current, 56
electron tube (see vacuum tube)
emitter, 56
emitter follower, 75, 275
envelope detector, 184
error amplifier, 204
error voltage, 209
external power, 38

F

F (class letter) 23
feedback, 77, 103
filament transformer, 110, 207
filter, low-pass, 146
filters, 223
first detector, 159
flasher (symbol), 29
flip-flop, 244, 263
fluctuating voltage, 194
FM, 144
FM discriminator, 178
free-running multivibrator, 132, 260

frequency comparator, 130, 270
frequency distortion, 104
frequency divider, 267
frequency doubler, 186, 270
frequency modulation, 144
frequency multiplication, 269
"front end," 86
fuse, 197
fuse (symbol), 23
fusible resistor, 199

G

gate, AND, 236, 243
  OR, 237, 243
  NAND, 241, 242
  NOR, 239
gated-beam FM detector, 181
generator, synchro, 255
graphic symbols, 17
Greek alphabet, 292
grid, 55
grid leak, 55, 80, 177
grid leak detecto 174
grid modulation 142
ground (symbol) 23
grounded-cathode circuit, 54, 78, 86, 91
grounded-collector circuit, 275
grounded-grid circuit, 91
grounded-plate circuit, cathode follower, 273

H

handset (symbol), 19
Hartley oscillator, 117
hash suppressor, 218
Hay bridge, 251
headset (symbol), 19
heptode, 160
heterodyning, 159
high-pass filter, 223
high-voltage power supply for cathode-ray tube, 208
high-voltage power supply for transmitter, 207

I

hole current, 56, 98
horizontal-output tube, 212
horizontal yoke coils, 212
H-pad, 222

I

IF amplifier, 86
IF transformer, 86
ignitron, 216
ignitron (symbol), 33
II, 74
inductive reactance, 115
inductors:
    adjustable (symbol), 24
    general (symbol), 23
    magnetic core (symbol), 23
    tapped (symbol), 24
industrial power controls, 218
inhibiting circuit, 275
integrated circuit, 74, 90
integrated circuit (symbol), 24
integrating circuit, 183
integration, 229
intermittent defects, 46, 49
International System of Units (SI units), 286
inverse current feedback, 105
inverse feedback, 103
inverse voltage feedback, 105
inverters, 215, 236
inverting input, 74
iron-core choke, 54
iron-core transformer, 54
I signal, 154

J

J (class letter), 22
jack (symbol), 22
JFET (symbol), 27
J-K flip-flop, 246
junction field-effect transistor (symbol), 27

K

K (class letter), 25
keyed AGC, 172

L

lamp(s) (symbols), 24
LC bridge, 249
L-C oscillators, 117
LED (symbol), 26
L-filter, 224
light-dimmer control circuit, 219
light-emitting diode, 273
light-emitting diode (symbol), 26
limit(er), 150, 178, 233
limiter grid, 182
linear RF amplifier, 112
linearity control, 214
line filter, 197
load resistor(s), 54, 78
local oscillator, 150, 158
logic circuits, 234
loop antenna (symbol), 18
loopstick antenna, 161
loudspeaker, 96
loudspeaker (symbol), 19
low-frequency compensation, 84
low-pass filter, 146, 223
L-pad, 221
LS (class letters), 19

M

M (class letter), 25
magnetic amplifier, 113
magnetic amplifier (symbol), 18
magnetic-core inductor (symbol), 23
master oscillator, 148
master, synchro, 255
Maxwell bridge, 251
M-derived filters, 224
meter (symbol), 25
metric system (SI units), 286

Index

microphone (symbol), 19
minimum-loss pads, 221
mixer, 147, 158
modifiers, 221
modulation, modulator, 140
  amplitude, 140
  Armstrong balanced, 147
  color, 152
  doubly balanced, 153
  frequency, 144
  grid, 142
  phase, 151
  plate, 141
monostable multivibrator, 261
MOS/FET, 281
MOS/FET amplifier, 282
motor, synchro, 255
multiplexing, 157
multivibrator, 132, 244, 260

N

NAND gate, 235, 241, 242, 243
negative feedback, 77, 81
neutralization, 108
neutralizing capacitor, 109
NII, 74
noninverting input, 74
NOR gate, 235, 239, 243

O

Ohm's Law formulas, 290
one-shot multivibrator, 261
O-pad, 222
op-amp, 72
open-loop gain, 77
operational amplifier, 72
OR gate, 235, 237
oscillators, 117
  Armstrong, 127
  blocking, 128
  Colpitts, 120
  crystal, 124, 125

oscillators (con't.)
  electron-coupled, 122
  Hartley, 117
  master, 148
  multivibrator, 132, 244, 261, 263
  relaxation, 137, 138
  sawtooth, 135, 266
  tuned-grid, 127
  tuned-plate, tuned-grid (TPTG), 123
output transformer, 96
overload protection, 204
Owen bridge, 251

P

P (class letter), 22
pads, 221
PAM, 156
parallel-resonant circuit, 83
part substitution, 49
passive elements, 38, 41
PCM, 156
PDM, 156
peak detector, 170
peaking coils, 78, 83
pentagrid converter, 38, 160
pentagrid tube, 160
pentode, 82
pentode (symbol), 32
permeability, 115
permeability tuned, 87
PFM, 156
phase angle, 152
phase inverter, phase splitter, 101, 106, 107
phase modulation, 151
photocells, 272
photoconductive transducer, 273
photoconductive transducer (symbol), 26
photoemissive diode, 273
photoemissive diode (symbol), 26
photosensitive diode (symbol), 26
photovoltaic transducer, 272
pickup head(s) (symbols), 30
piezoelectric crystal unit (symbol), 23

piezoelectric effect, 126
pi-filter, π-filter, pi (π)-filter, 38, 115, 192, 194, 197, 224
pilot light, 195
pi (π)-pad, 222
plate, 55
plate modulation, 141
PLM, 156
plug (symbol), 22
PM demodulation, 188
polarized capacitor (symbol), 21
positive feedback, 77, 118, 120
potentiometer (symbol), 26
power amplifiers:
    push-pull, 99
    single-ended, 96
power supplies, 191
    battery, 191
    bridge rectifier, 201
    full-wave, 195, 198, 201
    half-wave, 192
    high-voltage, 207
    regulated, 203
    transformerless voltage-doubler, 198
    transformerless voltage-tripler, 200
    universal AC-DC, 192
PPM, 156
pre-emphasis, pre-emphasized, 148, 150
preferred values, 285
professional troubleshooting, 42
PU (class letters), 30
pulse modulation (PM), 156
push-button switch (symbol), 28
push-pull, 99, 110
PWM, 156

Q

Q (class letter), 26
Q signal, 155
quadrature grid, 182
quartz crystal, 126
quartz crystal unit (symbol), 23

R

R (class letter), 25
radio-frequency spectrum, 293
ratio detector, 180
RC-coupled audio voltage amplifiers, 53
R-C oscillators, 117, 128
reactance circuit, 144
reactance, inductive, 115
rectifier, 192
    full-wave, 195
    half-wave, 192
regenerative detector, 176
regenerative feedback, 77
register, 246
regulation, 203
regulator, 253
relaxation oscillator, 137
relay(s) (symbols), 25
resistance checks, 50
resistor(s):
    adjustable contact (potentiometer), with (symbol), 26
    adjustable contact (rheostat), with (symbol), 26
    fusible, 199
    general (symbol), 25
    photoconductive transducer (symbol), 26
    surge, surgistor, 199
    tapped (symbol), 25
    thermistor (symbol), 26
resolver, 256
resolver (symbol), 29, 30
resonance bridge, 252
resonant circuit(s), 83, 107
RF amplifiers:
    linear, 112
    push-pull, 110
    single-ended, 107
rheostat (symbol), 26
rotary switches (symbols), 29
rotor winding, 254
RTL NOR gate, 239

S

S (class letter), 28
saturated, 115, 236
sawtooth generator, 135, 266
sawtooth signal, 128, 130
schematic analysis, 34
schematic diagram, purpose of, 17
Schering bridge, 252
Schmitt trigger, 263
SCR (symbol), 27
screen grid, 79, 89
second detector, 159
semiconductor devices:
    breakdown or Zener diode (symbol), 26
    diac, 219
    diac (symbol), 27
    diode, general (symbol), 26
    photocell, solar cell (symbol), 28
    photoemissive diode (symbol), 26
    photosensitive diode (symbol), 26
    phototransistor (symbol), 28
    silicon controlled rectifier, SCR, 216
    silicon controlled rectifier (symbol), 27
    thyristor, 216
    thyristor (symbol), 27
    transistor(s):
        insulated-gate field-effect (symbols), 27, 28
        junction field-effect (symbol), 27
        NPN (symbol), 27
        PNP (symbol), 27
        unijunction, UJT (symbol), 27
    triac, 219
    triac (symbol), 27
    tunnel diode (symbol), 26
series capacitance, 227
series clipper, 231
series peaking, 83
series regulator, 204
series resistors, 222
series-resonant circuit, 83
series-string filaments, 192

servomechanism, 253
shielded conductor (symbol), 20
shift register, 246
shunt clipper, 232
shunt inductance, 227
shunt peaking, 83
shunt resistors, 222
signal ground, 58
signal injection, 48
signal tracing, 46
silicon controlled rectifier, SCR (symbol), 27
single-shot multivibrator, 261
SI units, 286
slave, synchro, 255
S-R flip-flop, 245
stagger-tuned, 90
static tests, 45, 46, 50
stator winding, 253
steering circuit, 264
step-down transformer, 88
step-up transformer, 87
stereo adapter, 185
stereo broadcasting, 185
stereo-multiplex circuit, 184
superheterodyne, 86, 160
surge resistor, surgistor, 199
swinging choke, 207
switches, 191
switches (symbols), 28, 29
switching-bridge detector, 184
symmetrical pads, 221
synchro (symbols), 29
synchronized signal, synchronizing, sync, 117, 129, 135
synchronous detector, 187
synchro systems, 253
synchro types, 256

T

T (class letter), 31
tank circuit, 107

tapped inductor (symbol), 24
TDM, 156
telemetering, 157
thermistor, 102
thermistor (symbol), 26
thyratron, 215, 218
thyratron (symbol), 33
thyristor, 216
thyristor (symbol), 27
tickler coil, 164, 177
time-division multiplexing (TDM), 156, 157
time sharing, 157
tone controls, 60, 61
T-pad, 222
transformer:
  auto- (symbol), 32
  balun, 93
  coupling, 87
  filament, 110, 207
  general (symbol), 31
  IF, 86
  iron-core, 54
  magnetic-core (symbol), 31
  output, 96
  plate, 207
  saturating (symbol), 31
  shielded (symbol), 31
  step-down, 88
  step-up, 87
  tapped (symbol), 31
  turns ratio, 98
transistors:
  IGFET (symbol), 27, 28
  JFET (symbol), 27
  NPN (symbol), 27
  photo- (symbol), 28
  PNP (symbol), 27
  operation, 279
      unijunction (symbol), 27
treble tone control, 61
triac, 219
triac (symbol), 27
triac power-control circuit, 219
trimmer capacitor, 162
troubleshooting technique, general, 43
TTL NAND gate, 242

tuned-grid oscillator, 127
tuned-plate, tuned-grid (TPTG) oscillator, 123
tuner, TV, 92
tunnel diode (symbol), 26
turns ratio, 98
TV high-voltage supply, 211
twisted pair (symbol), 20
type number, 38

U

U (class letter), 24
unbalanced pads, 221
unijunction transistor, UJT (symbol), 27
unijunction transistor oscillator, 138
universal AC-DC power supply, 192
universal block diagram, 34
universal troubleshooting chart, 44
U-pad, 222

V

V (class letter), 32
vacuum tube(s):
  cathode-ray (symbol), 32
  gas, 215
  gas (symbol), 33
  pentode (symbol), 32
  picture tube(s) (symbols), 32, 33
  triodes (symbols), 32
variable capacitor (symbol), 21
variable L-pad, 223
variable T-pad, 223
V_{cc}, 65
vectors, 151
vibrator power supply, 216
video amplifier, 78
video detector, 169
visual inspection, 44, 45
voltage amplifiers, 52
voltage analysis, 50

Index

voltage dividers, 202
voltage regulation, 203
voltage reference, 204
voltage sampling divider, 204

width control, 214
Wien bridge, 252

W

Y

W (class letter), 20
wafer switch(es) (symbols), 29
waveform analysis, 48
waveguide (symbol), 21
Wheatstone bridge, 248

Y (class letter), 23
Y signal, 154

Z

Zener diode, 205
Zener diode (symbol), 26